Editorial Policy

§ 1. Lecture Notes aim to report new developments - quickly, informally, and at a high level. The texts should be reasonably self-contained and rounded off. Thus they may, and often will, present not only results of the author but also related work by other people. Furthermore, the manuscripts should provide sufficient motivation, examples and applications. This clearly distinguishes Lecture Notes manuscripts from journal articles which normally are very concise. Articles intended for a journal but too long to be accepted by most journals, usually do not have this "lecture notes" character. For similar reasons it is unusual for Ph. D. theses to be accepted for the Lecture Notes series.

§ 2. Manuscripts or plans for Lecture Notes volumes should be submitted (preferably in duplicate) either to one of the series editors or to Springer- Verlag, Heidelberg. These proposals are then refereed. A final decision concerning publication can only be made on the basis of the complete manuscript, but a preliminary decision can often be based on partial information: a fairly detailed outline describing the planned contents of each chapter, and an indication of the estimated length, a bibliography, and one or two sample chapters - or a first draft of the manuscript. The editors will try to make the preliminary decision as definite as they can on the basis of the available information.

§ 3. Final manuscripts should preferably be in English. They should contain at least 100 pages of scientific text and should include
- a table of contents;
- an informative introduction, perhaps with some historical remarks: it should be accessible to a reader not particularly familiar with the topic treated;
- a subject index: as a rule this is genuinely helpful for the reader.

Further remarks and relevant addresses at the back of this book.

Lecture Notes in Mathematics 1698

Editors:
A. Dold, Heidelberg
F. Takens, Groningen
B. Teissier, Paris

Springer
Berlin
Heidelberg
New York
Barcelona
Budapest
Hong Kong
London
Milan
Paris
Singapore
Tokyo

M. Bhattacharjee   D. Macpherson
R. G. Möller   P. M. Neumann

Notes on Infinite Permutation Groups

Springer

Authors

Meenaxi Bhattacharjee
Department of Mathematics
Indian Institute of Technology
Guwahati
Panbazar, Guwahati-781001, India
e-mail: meenaxi@iitg.ernet.in

Rögnvaldur G. Möller
Science Institute
University of Iceland
IS-107 Reykjavík, Iceland
e-mail: roggi@raunvis.hi.is

Dugald Macpherson
Department of Pure Mathematics
University of Leeds
Leeds LS2 9JT, UK
e-mail: pmthdm@amsta.leeds.ac.uk

Peter M. Neumann
The Queen's College
Oxford OX1 4AW, UK
e-mail: Peter.Neumann@queens.ox.ac.uk

Copyright © 1997 Hindustan Book Agency (India) and
Copyright © 1998 Springer-Verlag Berlin Heidelberg
Published in cooperation with Hindustan Book Agency (India), New Delhi
Outside India exclusive distribution rights remain with Springer-Verlag Berlin Heidelberg

Library of Congress Cataloging-in-Publication Data

Notes of infinite permutation groups / M. Bhattacharjee ... [et al.].
    p.   cm. -- (Lecture notes in mathematics, ISSN 0075-8434 :
1698)
  Includes bibliographical references and index.
  ISBN 3-540-64965-4 (softcover)
    1. Permutation groups.   I. Bhattacharjee, M. (Meenaxi), 1965-
II. Series: Lecture notes in mathematics (Springer-Verlag) ; 1698.
QA3.L28 no. 1698
[QA175]
510 s--dc21
[512.2]   98-39226
                                                            CIP

Mathematics Subject Classification (1991): 20B07, 20B15, 03C50

ISSN 0075-8434
ISBN 3-540-64965-4 Springer-Verlag Berlin Heidelberg New York

Typesetting: Camera-ready T$_E$X output by the author
SPIN: 10650132   41/3143-543210 - Printed on acid-free paper

Preface

The oldest part of group theory is that which deals with finite groups of permutations. One of the newest is the theory of infinite permutation groups. Much progress has been made during the last two decades, but much remains to be discovered. It is therefore an excellent research area: on the one hand there is plenty to be done; on the other hand techniques are becoming available. Some research takes as its goal the generalisation, extension or adaptation of classical results about finite permutation groups to the infinite case. But mostly the problems of interest in the finite and infinite contexts are quite different. In spite of its newness, the latter has already developed a momentum and an ethos of its own. It has also developed strong links with logic, especially with model theory.

To bring this lively and exciting subject to the attention of mathematicians in Assam—both students and established scholars—a course of sixteen lectures was presented in August and September 1996 at the Indian Institute of Technology, Guwahati. In the available time it was not possible to explore more than a fraction of the area. The lectures were therefore conceived with restricted aims: firstly, to expound a useful amount of general theory; secondly, to introduce and survey just one of the rich areas of recent research in which considerable progress has been made, namely, the theory of Jordan groups. That limited aim is reflected in this book. Although it is a considerable expansion of the lecture notes (and we have included many exercises, which range in difficulty from routine juggling of definitions to substantial recent research results), it is not intended to introduce the reader to more than a small—though important and interesting—part of the subject.

A course at this level can only succeed if it is a collaboration between lecturers and audience. That was so in this case and although the four speakers are those listed as authors of this volume, credit would be spread much more widely if title-page conventions permitted. In particular, we record our very warm thanks to the audience for being so enthusiastic and alert, and especially to the three note-takers Dr. B. K. Sharma, Ms. Shreemayee Bora and Ms. Shabeena Ahmed.

We are very grateful to Dr. S. Ponnusamy and Dr. K. S. Venkatesh for help with typesetting and with the figures. We wish to thank Prof. Moloy Dutta for help with typing and editing, and. Shabeena and Shreemayee, a second time, for help with proof-reading. We also thank the Editors of the TRIM series, specially Professors R. Bhatia and C. Musili, for accepting the book for publication in the series.

MB (Guwahati), HDM (Leeds), RGM (Reykjavík), IIMN (Oxford): April 1997

Some more acknowledgements:

The three of us who came from overseas have warm thanks to record to various institutions that provided support: to the Royal Society of London (Macpherson and Neumann); to the Indian National Science Academy (Neumann); to the Indian Institute of Technology Guwahati (Macpherson, Möller and Neumann) for a very kind welcome, for academic facilities and for help with travel within India. Our warmest thanks and congratulations, however, are reserved for our colleague and co-author Dr Meenaxi Bhattacharjee. She conceived and organised the whole project, she provided delightful company and hospitality in Assam, and, with characteristic energy she has brought this book project to a satisfactory conclusion.

HDM (Leeds), RGM (Reykjavík), IIMN (Oxford): April 1997

About the lecture course:

This book is based on lectures delivered by the four authors at the 'Lecture Course on Infinite Permutation Groups' that was held at the Indian Institute of Technology (IIT) Guwahati, India from 6th August to 19th September, 1996. The course was originally conceived as an informal series of lectures coinciding with the visits of Dr. Neumann, Dr. Macpherson and Dr. Möller to the Institute during that period.

The lectures were aimed at graduate students, research scholars, teachers and research workers of mathematics drawn from the colleges, universities, research institutes and other academic institutions in this region. The lectures (each lasting 90 minutes) were held twice a week. The course generated a tremendous amount of interest and as many as 60 participants attended the lectures.

The first speaker was Dr. Möller who introduced the participants to the basics of permutation group theory. I spoke next on wreath products of groups. In his lectures, Dr. Neumann explained the general theory of Jordan groups. Dr. Macpherson delivered the concluding lectures in which he developed the theory further, leading to the classification of infinite primitive Jordan groups. He also introduced the audience to the basic concepts of model theory, and illustrated the connections between model theory and Jordan groups using the Hrushovski construction. The sixteen chapters of this book correspond roughly to the sixteen lectures delivered at the course.

We wish to record our thanks to Professors M. S. Raghunathan and S. G. Dani of the National Board of Higher Mathematics for extending financial support, and to the faculty and staff of IIT Guwahati for their help in organising the course. We are particularly grateful to the Director of IIT Guwahati, Prof. D. N. Buragohain, for providing all infra-structural facilities and for his kind encouragement and support. We are also grateful to the Head of the Department of Mathematics at IIT Guwahati, Prof. P. Bhattacharyya, for giving us his whole-hearted help and guidance. On a personal note, my son Amlan deserves a special word of thanks for patiently tolerating my total preoccupation with the course last summer and then with the preparation of this book over the past few months.

I extend my warm and sincere thanks to each of my three co-authors—my teacher and *guru* Dr. Peter M. Neumann, my friend and collaborator Dr. Dugald Macpherson, and, my friend and *gurubhai* Dr. Rögnvaldur G. Möller—on my own behalf as well as on behalf of everyone else involved in the project, for accepting our invitation to visit Guwahati (and smilingly putting up with many odds during their Indian sojourn), for agreeing to speak at the course, for delivering such superb lectures and for being readily available to the participants for consultation at other times. By doing so they have given a rare opportunity to the participants to benefit from their expertise in the subject. Co-ordinating the course was both thrilling as well as challenging—and I have gained a lot from the experience.

MB (Guwahati): April 1997

Addresses of the authors:

Dr. Meenaxi Bhattacharjee
Department of Mathematics,
Indian Institute of Technology, Guwahati
Panbazar, Guwahati–781001
India
e-mail:
meenaxi@iitg.ernet.in

Dr. Rögnvaldur G. Möller
Science Institute,
University of Iceland
IS-107 Reykjavik
Iceland

roggi@raunvis.hi.is

Dr. Dugald Macpherson
Department of Pure Mathematics,
University of Leeds
Leeds LS2 9JT; England
e-mail:
pmthdm@amsta.leeds.ac.uk

Dr. Peter M. Neumann
The Queen's College
Oxford OX1 4AW
England

Peter.Neumann@queens.ox.ac.uk

Contents

Chapter 1

Some Group Theory

In this chapter we state (mostly without proof) some basic facts from the theory of groups. We have included only those facts from the general theory that we shall have occasion to use later – the treatment is by no means exhaustive. We refer to any standard textbook on Group Theory for more detailed accounts of the subject in general and for the contents of this chapter in particular. Some of the best are Hall (1959), Suzuki (1982, 1986), Rotman (1995), Robinson (1995) and Luthar & Passi (1996).

A *group* is a set together with a binary operation $\cdot : G \times G \to G$ and a distinguished element, called the *identity* element and usually denoted by 1, such that

(i) $a \cdot (b \cdot c) = (a \cdot b) \cdot c$ for all $a, b, c \in G$,

(ii) $1 \cdot a = a = a \cdot 1$ for all $a \in G$,

(iii)   for each $a \in G$ there exists a unique element a^{-1}, usually called the *inverse* of a in G, such that $a \cdot a^{-1} = 1 = a^{-1} \cdot a$.

The binary operation is usually called multiplication and we often write just ab instead of $a \cdot b$.

A group G is said to be *abelian* if $ab = ba$ for all $a, b \in G$.

A subset H of a group G is a *subgroup* of G if

(i) $1 \in H$,

(ii) $a \in H$ implies $a^{-1} \in H$,

(iii) $a, b \in H$ implies $ab \in H$.

To express that H is not merely a subset but a subgroup of G, we write $H \leq G$.

Lemma 1.1 *A subset H of a group G is a subgroup of G if and only if $H \neq \emptyset$ and $ab^{-1} \in H$ for all $a, b \in H$.*

The whole group G and the set $\{1\}$ containing only the identity element are trivially subgroups of any group G. A subgroup of G is said to be *proper* if it is not equal to G and *non-trivial* if it is not equal to $\{1\}$. We say that a proper subgroup H of G is a *maximal* subgroup of G if there is no subgroup properly containing H and properly contained in G.

Let $H \leq G$. For $a \in G$ define the *(right) coset* Ha by

$$Ha := \{ha \mid h \in H\}.$$

It is easy to see that either $Ha \cap Hb = \emptyset$ or $Ha = Hb$, and that the cosets of H form a partition of G. Two elements a and b of G are in the same coset of H if and only if $ab^{-1} \in H$. The number of cosets is known as the *index* of H in G and denoted by $|G : H|$. The set of cosets of H in G is denoted by G/H or by $\cos(G : H)$.

A subset $T = \{t_i\}$ of G is called a *(right) transversal* for H in G if $G = \bigcup_{t_i \in T} Ht_i$ and $Ht_i \neq Ht_j$ whenever $i \neq j$. Clearly the size of a transversal of H in G is equal to the index of H in G. Left cosets and left transversals are defined analogously.

Let $|S|$ denote the number of elements in the set S. Then we have the following theorem.

Theorem 1.2 (Lagrange's Theorem) *If G is a finite group then*

$$|G| = |G : H| \cdot |H|.$$

If G is a finite group then $|G|$ is called the *order* of the group G. Translated into this terminology, Lagrange's Theorem implies that *the order of a subgroup of a group divides the order of the group, when the group is finite.*

Let N be a subgroup of G. If $a^{-1}Na = N$ for every $a \in G$ then we say that N is a *normal* subgroup of G, and write $N \trianglelefteq G$. In this case, it is clear that $aN = Na$. The identity subgroup $\{1\}$ and the whole group G are trivially normal subgroups of G. If a group has no other normal subgroup then it is said to be a *simple* group.

For subsets A and B in G define

$$AB := \{ab \mid a \in A, b \in B\}.$$

If N is a normal subgroup of G then

$$(Na)(Nb) = N(aN)b = N(Na)b = N(ab),$$

so that we can define the product of the two cosets Na and Nb as Nab. It can be easily checked that this defines a group operation upon the set of cosets G/N. The group G/N is called the *quotient group* or the *factor group* of G by N.

Consider a chain

$$\{1\} = A_0 \trianglelefteq A_1 \trianglelefteq \ldots \trianglelefteq A_n = G \qquad (*)$$

of subgroups of a group G such that each A_i is a normal subgroup of A_{i+1}. The groups A_i are called *subnormal* subgroups of G, and such a chain of subgroups is called a *subnormal series*. If each A_i is normal in G then the series is called a *normal series*. Associated with every subnormal series is a sequence of factor groups A_{i+1}/A_i. A subnormal series in which each of these factor groups is simple is called a *composition series*.

EXERCISES:

1(i) Show that a subnormal series is a composition series if and only if each of the subgroups A_i, in the chain $(*)$, is a maximal normal subgroup of A_{i+1}.

1(ii) Two subnormal series are said to be *equivalent* if it is possible
to set up a one-one correspondence between the factors of the
two series such that the paired factors are isomorphic. Show
that any two composition series of a group are equivalent (this
is the famous Jordan–Hölder Theorem). Is the same true for
two normal series of a group?

A *homomorphism* from a group G to a group H is a mapping
$\theta : G \rightarrow H$ such that
$$(ab)\theta = (a\theta)(b\theta).$$

A surjective homomorphism is called an *epimorphism*. A homomorphism which is injective is called a *monomorphism*. A monomorphism
is sometimes also called an *embedding*. A bijective homomorphism is
called an *isomorphism*. When there is an isomorphism from G to H
then we say that G and H are *isomorphic*, and write $G \cong H$. An
isomorphism from G to itself is called an *automorphism*.

Lemma 1.3 *Suppose that* $\theta : G \rightarrow H$ *is a homomorphism between
the groups* G *and* H. *Then*

(i) $1\theta = 1$,

(ii) $(a^{-1})\theta = (a\theta)^{-1}$.

When $\theta : G \rightarrow H$ is a homomorphism we define the *image* of θ as

$$\operatorname{Im} \theta := \{g\theta \mid g \in G\},$$

and the *kernel* of θ as

$$\operatorname{Ker} \theta := \{g \in G \mid g\theta = 1\}.$$

Lemma 1.4 *With the above notation,* $\operatorname{Im} \theta$ *is a subgroup of* H *and*
$\operatorname{Ker} \theta$ *is a normal subgroup of* G.

Theorem 1.5 (First Isomorphism Theorem) *If* $\theta : G \rightarrow H$ *is a
homomorphism then*

Theorem 1.6 (Second Isomorphism Theorem) *If $H \leq G$ and $N \trianglelefteq G$ then $NH \leq G$, $H \cap N \trianglelefteq H$ and*

$$NH/N \cong H/(H \cap N).$$

Theorem 1.7 (Third Isomorphism Theorem) *If $N \trianglelefteq G$ and if $N \leq M \trianglelefteq G$, then $M/N \trianglelefteq G/N$ and*

$$(G/N)/(M/N) \cong G/M.$$

The set of all automorphisms of a group G forms a group called the *automorphism group* of G and is denoted by Aut (G). For a fixed $x \in G$, the map $G \to G$ which maps every element $g \in G$ to its conjugate $x^{-1}gx \in G$ is an automorphism of G, and is called the *inner automorphism* of G associated with the element x.

<u>EXERCISE:</u>

1(iii) Show that the set of all inner automorphisms of G forms a normal subgroup of Aut (G).

Given elements $x, g \in G$, the element $g^{-1}xg$ (sometimes written as x^g) is known as the *conjugate* of x by g and the set $\{g^{-1}xg \mid g \in G\}$ is called the *conjugacy class* of x in G. As an easy exercise in working with groups and elements of groups, let us prove the following lemma.

Lemma 1.8 *The set of conjugacy classes of elements in G forms a partition of G.*

Proof: We define a binary relation $\sim$ on G by declaring that $x \sim y$ if x is conjugate to y. In other words, $x \sim y$ if there exists $g \in G$ such that $x = y^g$. Then it is enough to show that $\sim$ is an equivalence relation. Clearly, it is reflexive, for $x = x^1$, so that $x \sim x$. If $x \sim y$ then $x = y^g = g^{-1}yg$ for some $g \in G$. But then $y = gxg^{-1} = x^{g^{-1}}$. Since $g^{-1} \in G$ we thus have that y is conjugate to x. Finally, if $x \sim y$ and $y \sim z$ then there exist $g, h \in G$ such that $x = g^{-1}yg$ and $y = h^{-1}zh$. But then $x = g^{-1}h^{-1}zhg = (hg)^{-1}z(hg) = z^{hg}$. Since $hg \in G$ we have $x \sim z$. This completes the proof of our claim. $\square$

Let H be a subgroup of G and $g \in G$. Then the set

$$H^g := g^{-1}Hg = \{g^{-1}hg \mid h \in H\}$$

is also a subgroup of G, and is said to be *conjugate* to the subgroup H. It is easy to check that the map $H \to H^g$ defined by $h \mapsto g^{-1}hg$ is a group isomorphism.

Exercise:

1(iv) Let G be a group and let $g \in G$. The *centraliser* of g in G is defined as
$$C_G(g) := \{h \in G \mid h^{-1}gh = g\}.$$

Show that $C_G(g)$ is a subgroup of G. Prove that the elements in the same conjugacy class have conjugate centralisers.

The *centre* of a group G, denoted by $Z(G)$. is the set of all $x \in G$ which commute with every element in G. Clearly $1 \in Z(G)$ and $Z(G) \trianglelefteq G$.

If A is a subset of a group G then the subgroup of G *generated* by A is the smallest subgroup of G containing A, and is denoted by $\langle A \rangle$. Alternatively, $\langle A \rangle$ is the set of those elements of G that can be written as products of elements in A and their inverses. If $\langle A \rangle = G$, then A is said to be a set of *generators* for G (or G is said to be *generated* by elements of A). A group generated by a single element is said to be *cyclic*.

Exercise:

1(v) Show that a cyclic group is abelian. Also show that a group of prime order is cyclic, and hence abelian.

1(vi) Show that an abelian group is simple if and only if it has prime order.

For $a, b \in G$, the element $a^{-1}b^{-1}ab$ of G is called the *commutator* of a and b. The *derived subgroup* or the *commutator subgroup* G' of a group G is the subgroup of G generated by all the commutators in G.

EXERCISE:

1(vii) (a) Show that $G' = \{1\}$ if and only if the group is abelian.
      (b) Show that $G' \lhd G$.
      (c) Prove that if $H \lhd G$, then the factor group G/H is abelian
      if and only if H contains the derived subgroup of G.

    The *(external) direct product* $A \times B$ of groups A and B is defined
on the cartesian product set $\{(a,b) \mid a \in A, b \in B\}$, with multiplication defined componentwise. That is, for all $a_1, a_2 \in A$ and $b_1, b_2 \in B$

$$(a_1, b_1)(a_2, b_2) = (a_1 a_2, b_1 b_2).$$

    It is often nice to be able to think of the group we are working
with as a direct product. If we can find subgroups A and B of a
group G such that

(i)   both A and B are normal subgroups of G, and

(ii) $G = AB$ and $A \cap B = 1$

then $A \times B \cong G$ and G is called the *(internal) direct product* of A
and B.

    Given an external direct product $A \times B$, we can identify the group
A with $\hat{A} := \{(a, 1) \mid a \in A\}$, and B with $\hat{B} := \{(1, b) \mid b \in B\}$. Then
it follows that the external direct product $A \times B$ is also the internal
direct product of its subgroups $\hat{A}$ and $\hat{B}$. The groups A and B are
called the *factors* of the direct product $A \times B$. Iterating the above
process, the direct product of a family of groups can also be defined.

    We can generalise the notion of a direct product to define a semi-
direct product of groups. A group G is an *(internal) semidirect pro-*
duct or a *split extension* of A by B, and denoted by $A \rtimes B$, if

(i) A is a normal subgroup of G and B is a subgroup of G, and

(ii) $G = AB$ and $A \cap B = 1$.

    Conversely, given two groups A and B and a homomorphism asso-
ciating an automorphism of A with every element of B, the *(external)*

semidirect product $A \rtimes B$ of A by B is also defined on the cartesian product set of A and B but with a slightly different definition for multiplication. For elements $a \in A$ and $b \in B$ let us denote the image of a under the automorphism of A associated with b by a^b. Then multiplication is given by the rule

$$(a_1, b_1)(a_2, b_2) = (a_1 a_2^{b_1^{-1}}, b_1 b_2).$$

By identifying the groups A and B with subgroups of $A \rtimes B$ as in the case of direct products, we can show that any internal semidirect product G is also an external semidirect product of groups, where the automorphism of A associated with the element $b \in B$ is the inner automorphism which takes an element $a \in A$ to the element $b^{-1}ab \in A$. Thus in practice we do not need to differentiate between external and internal (semi)direct products.

We end this chapter by defining a subdirect product of a family of groups. Let I be an index set and let $G := \prod_{i \in I} G_i$ be the direct product of a family of groups $\{G_i\}_{i \in I}$. An element $g \in G$ can then be expressed as $g = (g_i)_{i \in I}$ with $g_i \in G_i$. (Note that when the index set is infinite, the direct product defined above also includes elements for which infinitely many of the $g_i \neq 1$.) For each $i \in I$, let us define the *projection* $\pi_i : G \rightarrow G_i$ of G onto its i-th factor G_i by $g\pi_i = g_i$ if $g = (g_i)_{i \in I}$. A subgroup H of G is called a *subdirect product* of the family of groups $\{G_i\}_{i \in I}$ if for every $j \in I$ and for every $g_j \in G_j$ there exists an element $g \in H$ such that $g\pi_j = g_j$. In other words, H is a subdirect product of the family $\{G_i\}_{i \in I}$ if H projects *onto* each factor G_i of the direct product.

EXERCISE:

1(viii) Let $I := \{1, 2, \ldots, m\}$ and for every $i \in I$ let G_i be a simple group. If

$$H \leq G_1 \times G_2 \times \cdots \times G_m$$

is a subdirect product then show that

$$H \cong D_1 \times D_2 \times \cdots \times D_k,$$

with $k \leq m$ and where there exist distinct $i_1, i_2, \ldots, i_k \in I$ such that $D_j \cong G_{i_j}$ for each $j = 1, 2, \ldots, k$.

Chapter 2

Groups acting on Sets

A group is a permutation group if it has a faithful action on a set (cf. Sec. 2.2). And all groups can be considered to be permutation groups, usually in many different ways. Thus a study of these actions is crucial to the understanding of groups. For a detailed analysis of group actions, and for the theory of groups in terms of group actions see Neumann, Stoy & Thompson (1994). Any textbook on group theory, Hall (1959) for example, can be consulted for more information on the subject of permutations and symmetric groups. For a comprehensive account of permutation groups see Dixon & Mortimer (1996).

2.1 Group actions

Definition 2.1 Let G be a group and Ω a set. An *action* of G on Ω is a map $\Omega \times G \to \Omega$ written $(\omega, g) \mapsto \omega^g$ such that

(i) for every $g, h \in G$ and $\omega \in \Omega$, we have $(\omega^g)^h = \omega^{gh}$ and

(ii) for every $\omega \in \Omega$, we have $\omega^1 = \omega$, where 1 denotes the identity element of the group G.

If a group G has an action on Ω, we say that Ω is a *G-set* or a *G-space*.

Let us now see what this means in practice. To describe our simplest examples we need the following definitions.

Definition 2.2 A *permutation* of a set Ω is a bijective map $\Omega \to \Omega$. If g is a permutation then we write ω^g for the image of ω under g.

9

Definition 2.3 The set of all permutations on a set Ω is a group under the operation of composition of mappings. We call this group the *symmetric group* on Ω and denote it by $\mathrm{Sym}\,(\Omega)$.

If Ω is a finite set of size n, then $\mathrm{Sym}\,(\Omega)$ is also written as S_n.

Definition 2.4 A *permutation group* on a set Ω is a subgroup of $\mathrm{Sym}\,(\Omega)$.

Examples of group actions:

2(a) The group $\mathrm{Sym}\,(\Omega)$ has a natural action on Ω as each element of the symmetric group is a permutation on Ω. So has every permutation group on Ω. We shall see more of symmetric groups later in the chapter.

2(b) Let G be a group and set $\Omega := G$. Then G acts on Ω via right multiplication. That is, if $h \in \Omega$ and $g \in G$, then we set $h^g := hg$. This is called the *(right) regular representation* of G, or the *Cayley representation*. We shall explain the meaning of the terminology used here in the next chapter (cf. p. 22).

2(c) There is another natural way to make a group G act on itself. As in the last example, set $\Omega := G$ and define $h^g := g^{-1}hg$. It is easy to check that this defines an action on G called *conjugation*.

2(d) Let $\mathrm{GL}(2, \mathbb{R})$ be the group of all 2×2 invertible real matrices. Here the group operation is just the operation of matrix multiplication. This group has a natural action on $\mathbb{R}^2$ via right multiplication defined as follows: Given $M := \begin{pmatrix} a & b \\ c & d \end{pmatrix} \in \mathrm{GL}(2, \mathbb{R})$ and $(x, y) \in \mathbb{R}^2$ define

$$(x, y)^M := (x, y)M = (ax + cy, bx + dy).$$

To show that this defines a group action we only need to note that matrix multiplication is the same as composition of linear functions. We shall study matrix groups in detail in Chapter 7.

There are many other useful variations of the examples of group actions described above. For example, a group G acts on the set of

all its subgroups by conjugation. We discuss some more actions in
the exercises that follow.

EXERCISES:

2(i) There are many ways to construct a new group action from a
given group action. Let Ω be a G-space and let Γ be some set.
Define non-trivial G-actions on the sets:
(a) $\Omega \times \Omega$;
(b) $\Omega^{(k)} := \{(\omega_1,\ldots,\omega_k) \mid \omega_i \in \Omega \text{ and } \omega_i \neq \omega_j \text{ if } i \neq j\}$;
(c) $\Omega^{\{k\}}$, the set of all k-element subsets of Ω;
(d) $\wp(\Omega)$, the set of all subsets of Ω, and called the *power set*
of Ω;
(e) $\{f \mid f : \Omega \to \Gamma\}$;
(f) $\{f \mid f : \Gamma \to \Omega\}$.

2(ii) Let G be a group and N a normal subgroup of G. Show that
G acts on N by conjugation, and that this action defines a
homomorphism $G \to \text{Aut}(N)$. Furthermore, explain that if
A is an abelian normal subgroup of G the same idea can be
used to define an action of G/A on A and a homomorphism
$G/A \to \text{Aut}(A)$.

2.2   Permutation groups

Let us have a closer look at actions of permutation groups. We have
already seen that a permutation group on Ω acts on Ω. But a group
acting on Ω is not necessarily a permutation group, although it very
nearly is, as the following analysis will show.

Let Ω be a G-space. We take $g \in G$ and define a map $\rho_g : \Omega \to \Omega$
by $\omega \rho_g = \omega^g$, for all $\omega \in \Omega$. Then

$$\omega \rho_g \rho_{g^{-1}} = (\omega^g)^{g^{-1}} = \omega^{gg^{-1}} = \omega^1 = \omega.$$

Similarly, $\omega \rho_g^{-1} \rho_g = \omega$. This shows that the map ρ_g is a permutation
on Ω, since it has both left and right inverses.

Thus we have a map $\rho : G \to \text{Sym}(\Omega)$ sending $g \in G$ to the
permutation $\rho_g \in \text{Sym}(\Omega)$. This map is a homomorphism because

for every $\omega \in \Omega$, we have

$$\omega \rho_{gh} = \omega^{gh} = (\omega^g)^h = (\omega \rho_g)\rho_h = \omega \rho_g \rho_h$$

so that $\rho_{gh} = \rho_g \rho_h$.

The kernel of this homomorphism ρ is

$$K := \{g \in G \mid \omega^g = \omega \text{ for all } \omega \in \Omega\}.$$

If $K = \{1\}$ then we say that the action of G is *faithful*, and then

$$G \cong \operatorname{Im} \rho \leq \operatorname{Sym}(\Omega).$$

In this case we can think of G as a permutation group on Ω. Otherwise G is a permutation group modulo the kernel K.

Definition 2.5 The homomorphism $\rho : G \to \operatorname{Sym}(\Omega)$ defined above is called the *permutation representation* of G on Ω and the image of G under ρ is a permutation group, denoted G^Ω, and called the permutation group *induced* by G on Ω.

EXERCISES:

2(iii) Show that the action on a group G by conjugation as defined in Example 2(c) is faithful if only if the group is centreless (that is, $Z(G) = \{1\}$).

2(iv) Show that there is a one-one correspondence between actions of a group G on a set Ω and representations of G by permutations of Ω.

Let us next describe another very useful way of manufacturing group actions. This is an extension of the idea behind the right regular representation described earlier.

Another example of a group action:

2(e) Let H be a subgroup of G. Recall that $\cos(G : H)$ is the set of right cosets of H in G. The group G then has a natural action on $\cos(G : H)$ given by $(Ha)^g := Hag$ for all $g \in G$. Observe that the (right) regular representation (cf. Eg. 2(b)) is a special case where $H = \{1\}$.

Lemma 2.6 *The kernel K of the action of G on $\cos(G : H)$ is equal to $\bigcap_{x \in G} x^{-1} H x$, and is the biggest normal subgroup of G contained in H.*

Note here that $x^{-1} H x = \{x^{-1} h x \mid h \in H\}$, and the biggest normal subgroup of G contained in H means that every normal subgroup of G that is contained in H is contained in K. The group K is often called the *core* of H in G.

Proof: Suppose $a \in K$ and $x \in G$. Then $Hx = (Hx)^a = Hxa$. So $xax^{-1} \in H$ and hence $a \in x^{-1} H x$. Therefore $K \subseteq \bigcap_{x \in G} x^{-1} H x$. On the other hand, suppose $a \in \bigcap_{x \in G} x^{-1} H x$. Then for every element $x \in G$ there is some element $h_x \in H$ such that $a = x^{-1} h_x x$. Then $(Hx)^a = Hxa = Hx(x^{-1} h_x x) = H h_x x = Hx$. So $a \in K$. Thus $K = \bigcap_{x \in G} x^{-1} H x$ and the first part is done.

To prove the second part we let L be a normal subgroup of G and assume $L \leq H$. Then, for any $x \in G$, we have $x^{-1} L x \leq x^{-1} H x$. But $L \trianglelefteq G$ and so $x^{-1} L x = L$. Thus $L \leq x^{-1} H x$, for every $x \in G$ and therefore $L \leq \bigcap_{x \in G} x^{-1} H x = K$. $\square$

Corollary 2.7 *Let G be a group and $H \leq G$ with $|G : H| = n$, where n is finite. Then G has a normal subgroup K such that $K \leq H$ and $|G : K|$ divides $n!$.*

Proof: As we have seen in Example 2(e), the group G acts on the set $\cos(G : H)$ which has n elements. Therefore we have a homomorphism

$$\theta : G \rightarrow \text{Sym}\,(\cos(G : H)) = S_n.$$

Let K be the kernel of this homomorphism. We know that K is normal in G and $K \leq H$. By the First Isomorphism Theorem (Theorem 1.5), we have $G/K \cong \text{Im}\,\theta \leq S_n$. Thus by Lagrange's Theorem (Theorem 1.2), it follows that $|\text{Im}\,\theta|$ divides $|S_n| = n!$. This completes the proof as $|G : K| = |G/K| = |\text{Im}\,\theta|$. $\square$

The above corollary is an excellent illustration of the way in which group actions may be used to obtain results in the general theory of groups. We end this section by stating a refinement of the above result as an exercise.

<u>EXERCISE:</u>

2(v) Show that if G is simple and has a subgroup of index n, where
$n > 2$, then $|G|$ divides $n!/2$. [Note: The proof will need an
understanding of even permutations and the alternating group,
which we define in Sec. 2.4.]

2.3   Cycles and cycle types

Symmetric groups are central to the study of permutation groups
since every permutation group is a subgroup of a symmetric group of
suitable size. We will study some basic properties of these groups in
the following two sections.

As we have already seen in Section 2.1 a *permutation* of a set Ω
is a bijective map $\Omega \to \Omega$. We write ω^g for the image of ω under a
permutation g.

There are two common ways of writing down a permutation g on
a set Ω. The first is to write out all the elements of Ω in a row and to
write the image of each element under g directly underneath it. For
example, if Ω is the set $\{1, 2, 3, 4, 5\}$, and $1^g = 1, 2^g = 3, 3^g = 2, 4^g = 5$
and $5^g = 4$ then g is written as

$$g = \begin{pmatrix} 12345 \\ 13254 \end{pmatrix}$$

The second is to write the permutation in cycle notation. A *cycle* can
be either finite or infinite. A *finite cycle* of length n (or an n-cycle) is
a permutation $h = (\alpha_1 \ \alpha_2 \ \ldots \ \alpha_n)$ such that $\alpha_i^h = \alpha_{i+1}$ for $i < n$ and
$\alpha_n^h = \alpha_1$ and such that h fixes all the other points. For example, if g
is the permutation described above, we can write g in cycle notation
as $(1)(2\,3)(4\,5)$ or simply as $(2\,3)(4\,5)$ by omitting the fixed point. A
2-cycle is called a *transposition*. If Ω is infinite, an *infinite cycle* can
be defined similarly as a permutation $k = (\ldots \ \alpha_{-1} \ \alpha_0 \ \alpha_1 \ \ldots)$ on Ω
such that $\alpha_i^k = \alpha_{i+1}$ for all $i \in \mathbf{Z}$. Two cycles are said to be disjoint
if they have no element in common. To write a permutation in cycle
notation, we use the fact that every permutation, whether on a finite or an infinite set, can be expressed as a product of disjoint cycles.

We have already noted that the set of all permutations on a set Ω is a group under the operation of composition of mappings. It is called the *symmetric group* on Ω, and is denoted by $\mathrm{Sym}\,(\Omega)$. The size of Ω is said to be the *degree* of the symmetric group $\mathrm{Sym}\,(\Omega)$. If $|\Omega| = n < \infty$, we write $S_n := \mathrm{Sym}\,(\Omega)$ and call S_n the symmetric group of degree n.

Definition 2.8 Let $g \in \mathrm{Sym}\,(\Omega)$. For $i \in \mathbf{N}$ define a_i as the (cardinal) number of i-cycles that g has and let a_0 be the number of infinite cycles in g. Then the *cycle type* of g is defined to be

$$\aleph_0^{a_0} 1^{a_1} 2^{a_2} \ldots k^{a_k} \ldots .$$

Clearly, when Ω is finite, we must have $a_0 = 0$. Also, the cardinal sum $\aleph_0 a_0 + a_1 + 2a_2 + \ldots + ka_k + \ldots$ is equal to $|\Omega|$. We note that cycle types also take into account the number of fixed points in a permutation. When there is no scope for confusion, we will omit cycle lengths of exponent zero from the notation for a cycle type.

Examples :

- If $g = (1)(2\ 3)(4\ 5\ 6) \in S_6$ then the cycle type of g is $1^1 2^1 3^1$.

- If $g = (\ldots -1\ 0\ 1\ 2\ \ldots) \in \mathrm{Sym}\,(\mathbf{Z})$ then the cycle type of g is $\aleph_0^1 1^0 2^0 \ldots .$

EXERCISE:

2(vi) (Based on a problem from the International Mathematical Olympiad 1990.) Let $\mathbf{Q}^+$ denote the set of all positive rational numbers. A function $f : \mathbf{Q}^+ \to \mathbf{Q}^+$ satisfies the equation

$$f(x f(y)) = \frac{f(x)}{y} \text{ for all } x \text{ and } y \text{ in } \mathbf{Q}^+.$$

Prove that f must be a permutation of the set $\mathbf{Q}^+$ and must indeed be an automorphism of the multiplicative group $\mathbf{Q}^+$. What is the cycle type of f? Construct such a function.

Theorem 2.9 *Two permutations* $g, h \in \mathrm{Sym}\,(\Omega)$ *are conjugate in* $\mathrm{Sym}\,(\Omega)$ *if and only if g and h have the same cycle type.*

Proof: Let us first prove that for $n > 1$ if we conjugate an n-cycle $h := (\alpha_1 \ \alpha_2 \ \ldots \ \alpha_n)$ by an element f of $\mathrm{Sym}\,(\Omega)$, we get another n-cycle $(\alpha_1^f \ \alpha_2^f \ \ldots \ \alpha_n^f)$. This is because $\alpha_i^{f(f^{-1}hf)} = \alpha_i^{hf} = \alpha_{i+1}^f$ identifying α_{n+1} with α_1. The same holds for fixed points and for infinite cycles. Extending this result to products of disjoint cycles, it follows that if g and h are conjugate they have the same cycle type.

Conversely, suppose g and h have the same cycle type. We pair the cycles of g with the cycles of h in such a way that a k-cycle of g is paired with a k-cycle of h. Note that we must do this for the 1-cycles, i.e. fixed points, as well. Then we define a permutation f of Ω such that it maps the elements in a cycle of h to the elements in the paired cycle of g and preserves the cyclic order. We claim that $g = f^{-1}hf$.

To prove our claim, suppose that $(\alpha_1 \ \alpha_2 \ \ldots \ \alpha_k)$ is a k-cycle of g and let $(\beta_1 \ \beta_2 \ \ldots \ \beta_k)$ be the paired cycle of h. Let the permutation f be defined such that it maps β_i to α_i for $1 \le i \le k$. Then we get $\alpha_i^{f^{-1}hf} = \beta_i^{hf} = \beta_{i+1}^f = \alpha_{i+1} = \alpha_i^g$ identifying α_{k+1} with α_1. So the claim is true if g and h each consist of a single k-cycle alone. A similar argument applies to infinite cycles as well. But since the cycles in g (and also h) are disjoint we can do the same thing for all other cycles in g to extend f without affecting what we have already defined of f. This completes the proof of the theorem. $\square$

2.4   Basic facts about symmetric groups

Recall that the symmetric group on a set Ω is also denoted as S_n if Ω is a finite set of size n.

Theorem 2.10 *S_n is generated by transpositions.*

Proof: Any element in S_n can be written as a product of disjoint cycles. Therefore it is enough to show that any finite k-cycle (for $k > 2$) can be written as a product of transpositions. But this is easy because $(\alpha_1 \alpha_2 \ldots \alpha_k) = (\alpha_k \alpha_{k-1})(\alpha_{k-1} \alpha_{k-2})\ldots(\alpha_3 \alpha_2)(\alpha_2 \alpha_1)$. $\square$

Remark: We do not need all the transpositions to generate S_n. For example, the tranpositions $(1 \ 2), (2 \ 3), \ldots, (n-1 \ n)$ are enough to generate S_n. This can be seen as follows. If $(i \ j)$ is a transposition

in S_n, we can suppose that $i < j$ and that $j - i = k > 0$. Then
$(i\ j) = (i\ i+1)(i+1\ i+2)\ldots(i+k-2\ i+k-1)(i+k-1\ i+k)$
$(i+k-1\ i+k-2)\ldots(i+1\ i)$.

From the above theorem it is clear that any permutation in S_n can be expressed as a product of transpositions (although there might be more than one way of doing so).

Definition 2.11 A permutation $g \in S_n$ is *even* if it can be written as a product of an even number of transpositions and *odd* if it can be written as a product of an odd number of transpositions.

The following easy facts are left as exercises.

EXERCISES:

2(vii) A permutation is either odd or even but not both. [Hint: Consider the induced action of a permutation $g \in S_n$ on a polynomial $\prod_{i<j}(x_i - x_j)$, $i, j \in \{1, 2, \ldots, n\}$ which maps it to the polynomial $\prod_{i<j}(x_{ig} - x_{jg})$. Show that these two polynomials can differ only in sign.]

2(viii) A $(2k + 1)$-cycle is even and a $2k$-cycle is odd.

Definition 2.12 The *alternating group* A_n of degree n is the collection of all even permutations in S_n. That is

$$A_n := .\{g \in S_n \mid g \text{ is even}\}.$$

It is easy to prove that A_n is indeed a group. Furthermore, it is a normal subgroup of S_n with $|S_n : A_n| = 2$. The most important fact about the finite alternating groups is stated in the following theorem.

Theorem 2.13 *For $n \neq 4$, the group A_n is simple.*

One way to see this is by proving the following three claims:

(1) the group A_n is generated by 3-cycles;

(2) any two 3-cycles are conjugate in A_n;

(3) any proper normal subgroup of A_n must contain a 3-cycle.

We shall prove only the first as it corresponds to Theorem 2.10 which we have already done for symmetric groups.

Theorem 2.14 *The group A_n is generated by 3-cycles.*

Proof: By Theorem 2.10, we know that every element of S_n and hence also of A_n is a product of transpositions. Since every permutation in A_n is even, we know that the number of such transpositions in any such expression is even. Hence any such element can also be expressed as a product of pairs of transpositions. If the two transpositions in a pair are identical then they are equivalent to the identity permutation. If they have one letter in common, say $(a\ b)(a\ c)$ then their product is the 3-cycle $(a\ b\ c)$. Finally, if they have no element in common, say $(a\ b)(c\ d)$ then we insert a pair of identical transpositions between them to get two 3-cycles. Thus $(a\ b)(c\ d) = (a\ b)(a\ c)(a\ c)(c\ d) = (a\ b\ c)(a\ d\ c)$. This completes the proof. $\square$

As with symmetric groups, we do not need all the 3-cycles to generate A_n, much smaller sets suffice. For example, it can be shown that the 3-cycles $(1\ 2\ 3), (1\ 2\ 4), (1\ 2\ 5), \ldots, (1\ 2\ n-1), (1\ 2\ n)$ are enough. We shall see more of these groups in Chapter 6.

EXERCISES:

2(ix) Set $\Omega := \{1, \ldots, n\}$ and let S_n act on Ω in the usual manner. Show that
(a) the transpositions $(1\ 2), (1\ 3), \ldots, (1\ n)$ generate S_n;
(b) the permutations $(1\ 2)$ and $(1\ 2\ \ldots\ n)$ generate S_n.

2(x) Let X be the set of permutations in S_n that can be expressed as a product of r transpositions, where r is a multiple of 3. Show that $X = S_n$.

2(xi) The *order* of an element g in a group G is defined as the least finite number n such that $g^n = 1$. If no such finite n exists then g is said to be of infinite order. How can one determine the order of a permutation from its cycle type?

2(xii) Show that A_4 is not simple by finding a normal subgroup of A_4 with 4 elements.

Chapter 3

Transitivity

Most of group theory, and in particular the theory of group actions, is about symmetries. A natural question to ask therefore is whether two points can look the same. Translated into our language of group actions, this question can be stated as follows. Given two points α, β in a G-space Ω, can one find an element $g \in G$ such that $\alpha^g = \beta$? In this chapter, we formalise this question and discuss the immediate consequences if the answer to the above question is in the affirmative. For further reading, the reader is again referred to Wielandt (1960, 1964).

3.1 Orbits and transitivity

Definition 3.1 The *orbit* of an element α in Ω is the set

$$\alpha^G := \{\alpha^g \mid g \in G\}.$$

When $\alpha^G = \Omega$ then we say that G acts *transitively* on Ω, or that Ω is a *transitive G-space*.

Therefore, a group G is transitive if all elements of Ω lie in one orbit. Equivalently, G is transitive if and only if, for any distinct $\alpha, \beta \in \Omega$, there is always some $g \in G$ such that $\alpha^g = \beta$.

Define a binary relation on the G-space Ω by declaring that α 'is in the same orbit as' β if there is an element $g \in G$ such that $\alpha^g = \beta$. The following theorem will show that this is an equivalence relation. The equivalence classes are called the *orbits* of G in Ω.

19

Theorem 3.2 *Every G-space Ω can be expressed in a unique way as a disjoint union of orbits. Equivalently, the binary relation 'is in the same orbit as' defined above is an equivalence relation.*

Proof: We first show that either $\alpha^G = \beta^G$ or $\alpha^G \cap \beta^G = \emptyset$. For, suppose $\gamma \in \alpha^G \cap \beta^G$. Then $\gamma = \alpha^{g_1} = \beta^{g_2}$ for some $g_1, g_2 \in G$, and therefore $\alpha = \beta^{g_2 g_1^{-1}}$. So, $\alpha \in \beta^G$ and we can conclude that $\alpha^G \subseteq \beta^G$. In the same way, we can show that $\beta^G \subseteq \alpha^G$. Therefore, if $\alpha^G \cap \beta^G \neq \emptyset$, then $\alpha^G = \beta^G$.

Clearly, every element of Ω is contained in some orbit of G (because $\alpha \in \alpha^G$ for every $\alpha \in \Omega$), so our proof is complete. $\square$

If a subset Δ of Ω is a union of orbits, then we have a natural action of G on Δ. So we can consider Δ as a G-space in its own right. In the particular case when Δ is a single orbit, it is a transitive G-space. The group G induces a permutation group on Δ consisting of the set of restrictions to Δ of elements of G. This is denoted G^Δ and called a *transitive constituent* of G (see also remark after Definition 3.12). It is easy to see that every permutation group G is a subgroup of the cartesian product of its transitive constituents. It is, in fact, a subdirect product of them (cf. Ch. 1). Therefore, one can often initially restrict attention to transitive G-spaces and then obtain results for non-transitive G-spaces. Most of the G-spaces that occur in our discussion, however, will be transitive.

Examples :

3(a) Any set Ω as a $\mathrm{Sym}\,(\Omega)$-space is transitive.

3(b) The group $\mathrm{Sym}\,(\Omega)$ also acts on $\wp(\Omega)$, the power set of Ω, such that for $\Delta \subseteq \Omega$ and $g \in \mathrm{Sym}\,(\Omega)$ we have $\Delta^g := \{\delta^g \mid \delta \in \Delta\}$. This action, however, is not transitive (except when $\Omega = \emptyset$) because subsets of Ω of different cardinalities cannot be mapped to each other.

3(c) The set G itself as a G-space with the right regular representation is clearly transitive. More generally, given a subgroup H of G, the coset space $\cos(G : H)$ is also a transitive G-space for, given two cosets Ha and Hb, the element $a^{-1}b \in G$ takes Ha to Hb. We shall show in Theorem 3.6 that transitive G-spaces look exactly like coset spaces.

3(d) The group G acts on itself by conjugation. This action, however, is not transitive (unless $G = \{1\}$), because the identity element cannot be mapped to a non-identity element of G by conjugation. The orbits of G under this action are precisely the conjugacy classes of G.

3(e) The action of $\mathrm{GL}(2, \mathbf{R})$ on $\mathbf{R}^2$ is not transitive. This is because the zero element $(0, 0)$ of $\mathbf{R}^2$ cannot be moved to a non-zero element by a nonsingular matrix. Note, however, that all the non-zero elements of $\mathbf{R}^2$ lie in one $\mathrm{GL}(2, \mathbf{R})$-orbit. This is because given any two non-zero elements (x, y) and (u, v) in $\mathbf{R}^2$ the matrix

$$M := \begin{pmatrix} a & b \\ c & d \end{pmatrix}$$

where $a, b, c, d \in \mathbf{R}$ are such that $ax + cy = u$ and $bx + dy = v$ will map (x, y) to (u, v). We can easily arrange that M belongs to $\mathrm{GL}(2, \mathbf{R})$. So there are exactly two orbits of $\mathbf{R}^2$, namely $\{(0, 0)\}$ and $\mathbf{R}^2 \setminus \{(0, 0)\}$, as a $\mathrm{GL}(2, \mathbf{R})$-space.

3.2 Stabilisers and transitivity

Definition 3.3 For $\alpha \in \Omega$ we define the *stabiliser* of α in G to be the set

$$G_\alpha := \{g \in G \mid \alpha^g = \alpha\}.$$

It is easy to see that G_α is a subgroup of G. The stabiliser G_α is sometimes also called a *point stabiliser*.

Lemma 3.4 *For $g, h \in G$ we have that*

$$G_\alpha g = G_\alpha h$$

if and only if $\alpha^g = \alpha^h$.

Proof: We have

$$G_\alpha g = G_\alpha h \Leftrightarrow gh^{-1} \in G_\alpha \Leftrightarrow \alpha^{gh^{-1}} = \alpha \Leftrightarrow \alpha^g = \alpha^h. \quad \square$$

Definition 3.5 A map $\theta : \Omega \to \Omega'$ between two G-spaces Ω and Ω' is a G-morphism if for all $\omega \in \Omega$ and $g \in G$,

$$(\omega^g)\theta = (\omega\theta)^g.$$

If θ is bijective then it is called a G-isomorphism.

The following theorem will prove that transitive G-spaces look exactly like coset spaces.

Theorem 3.6 *Let Ω be a transitive G-space. Then Ω is G-isomorphic to the G-space $\cos(G : G_\alpha)$ for any $\alpha \in \Omega$.*

Proof: Define a map $\theta : \Omega \to \cos(G : G_\alpha)$ by $\omega\theta = G_\alpha g$ where $g \in G$ is such that $\alpha^g = \omega$. Note that such an element g exists because Ω is transitive. To show that θ is well-defined, let $h \in G$ be such that $\alpha^h = \omega$. Then $\alpha^g = \alpha^h$ so that using Lemma 3.4 we have $G_\alpha g = G_\alpha h$. Using the same lemma yet again we also get injectivity. Surjectivity is obvious. Thus θ is a bijection.

It only remains to show that θ is a G-morphism. That is, we have to show that for any $\omega \in \Omega$ and $g \in G$ we have $(\omega^g)\theta = (\omega\theta)^g$. To do this, find $h \in G$ such that $\alpha^h = \omega$. Then

$$(\omega^g)\theta = (\alpha^{hg})\theta = G_\alpha hg = (\omega\theta)^g. \quad \square$$

Corollary 3.7 *Under the hypotheses of the last theorem,*

$$|G : G_\alpha| = |\cos(G : G_\alpha)| = |\Omega|. \quad \square$$

Definition 3.8 A group G is said to act *regularly* on a set Ω if it is transitive and the stabiliser of every point in Ω is the identity.

By Theorem 3.6, if the action of G is regular then the action of G on Ω is isomorphic to its action on itself (and G acts on itself by right multiplication). This is therefore called the *right regular representation* of G (cf. Eg. 2(b)). Clearly, in this situation, G also acts faithfully on Ω. This action was used by Cayley to show that every group is isomorphic to a permutation group. Thus it is also called the *Cayley representation*.

We leave the easy proof of the following lemma about actions of transitive abelian groups as an exercise.

Lemma 3.9 *Any transitive abelian permutation group G (acting on a set Ω) is regular.* □

Transitivity also means that all the elements in Ω are the "same" with respect to the action of G. The tool we use to give this statement a definite meaning is conjugation.

Lemma 3.10 *Let $g \in G$ be such that $\alpha^g = \beta$. Then $g^{-1}G_\alpha g = G_\beta$.*

Proof: Recall that $g^{-1}Hg = \{g^{-1}hg \mid h \in H\}$. Take $h \in G_\alpha$. Then

$$\beta^{g^{-1}hg} = \alpha^{g(g^{-1}hg)} = \alpha^{hg} = \alpha^g = \beta$$

so that $g^{-1}G_\alpha g \leq G_\beta$. Conversely, let $k \in G_\beta$. Then by a similar argument, $gkg^{-1} \in G_\alpha$ and therefore $k \in g^{-1}G_\alpha g$ which gives the reverse inclusion. □

3.3 Extensions of transitivity

Definition 3.11 Let $k \in \mathbf{N}$. We say that a G-space Ω is *k-transitive* or that G acts *k-transitively* on Ω if for any two sets of k distinct points in Ω, say $\alpha_1, \alpha_2, \ldots, \alpha_k$ and $\beta_1, \beta_2, \ldots, \beta_k$ there is an element $g \in G$ such that $\alpha_i^g = \beta_i$ for $i = 1, 2, \ldots, k$.

A 1-transitive space is simply a *transitive space* as defined earlier. A 2-transitive space is often said to be *doubly transitive*. When Ω is infinite and is k-transitive for every $k \in \mathbf{N}$, it is said to be *highly transitive*.

We also need to extend the idea of stabilisers.

Definition 3.12 Let Ω be a G-space and $\Delta \subseteq \Omega$. We define the *setwise stabiliser* of Δ in G to be

$$G_{\{\Delta\}} := \{g \in G \mid \Delta^g = \Delta\}$$

and the *pointwise stabiliser* of Δ in G to be

$$G_{(\Delta)} := \{g \in G \mid \delta^g = \delta \text{ for all } \delta \in \Delta\}.$$

Clearly both $G_{\{\Delta\}}$ and $G_{(\Delta)}$ are subgroups of G. They coincide when Δ is a singleton set and are equal to the (point) stabiliser of that point in G. Also $G_{(\Delta)} \trianglelefteq G_{\{\Delta\}}$. We define G^{Δ} to be the factor group $G_{\{\Delta\}}/G_{(\Delta)}$, namely the permutation group induced by $G_{\{\Delta\}}$ on Δ.

Theorem 3.13 *Let Ω be a transitive G-space and $\alpha \in \Omega$. Then G acts $(k+1)$-transitively on Ω if and only if G_α acts k-transitively on $\Omega \setminus \{\alpha\}$. Furthermore, if G is k-transitive and $\Delta \subseteq \Omega$, then $G_{(\Delta)}$ is $(k-d)$-transitive on $\Omega \setminus \Delta$, where $d = |\Delta|$.*

Proof: If G is $(k+1)$-transitive, then we can always find $g \in G$ fixing α and moving a set of k distinct points to any other such set. Thus G_α is k-transitive on $\Omega \setminus \{\alpha\}$. To prove the converse, let $\alpha_1, \alpha_2, \ldots, \alpha_{k+1}$ and $\beta_1, \beta_2, \ldots, \beta_{k+1}$ be two sets of $(k+1)$ distinct points of Ω. By transitivity of Ω, there is some $g_1 \in G$ such that $\alpha_1^{g_1} = \beta_1$. Also let $\gamma_i := \alpha_i^{g_1}$ for $2 \leq i \leq k+1$. Then by our hypothesis, since G_{β_1} is k-transitive on $\Omega \setminus \{\beta_1\}$ there must exist $g_2 \in G_{\beta_1}$ which maps γ_i to β_i for $2 \leq i \leq k+1$. Then it is easy to check that $g_1 g_2$ maps α_i to β_i for $1 \leq i \leq k+1$. This completes the proof of the first part of the theorem.

To prove the second part, let $\alpha_1, \alpha_2, \ldots, \alpha_{k-d}$ and $\beta_1, \beta_2, \ldots, \beta_{k-d}$ be two sets of distinct points in $\Omega \setminus \Delta$ and let $\alpha_{k-d+1}, \alpha_{k-d+2}, \ldots, \alpha_k$ be the d distinct points of Δ. Since G is k-transitive we can find $g \in G$ which moves the set $\alpha_1, \alpha_2, \ldots, \alpha_k$ to $\beta_1, \beta_2, \ldots, \beta_{k-d}, \alpha_{k-d+1}, \ldots, \alpha_k$. This element g is clearly in $G_{(\Delta)}$. This completes the proof of the theorem. $\square$

Corollary 3.14 *Let Ω be a k-transitive G-space. Suppose $\Gamma, \Delta \subseteq \Omega$ such that $|\Gamma| = |\Delta| = k$. Then there exists some $g \in G$ such that $\Gamma^g = \Delta$ and such that $G_{(\Delta)} = g^{-1} G_{(\Gamma)} g$.*

Note that Γ^g denotes the set $\{\omega \in \Omega \mid \omega = \gamma^g, \text{ for some } \gamma \in \Gamma\}$.

Proof: The first part is immediate from the fact that G is k-transitive. Let $g \in G$ be such that $\Gamma^g = \Delta$. Let $h \in G_{(\Gamma)}$ and $\delta \in \Delta$. Then there exists $\gamma \in \Gamma$ such that $\gamma^g = \delta$. Now,

$$\delta^{g^{-1}hg} = \gamma^{g(g^{-1}hg)} = \gamma^{hg} = \gamma^g = \delta$$

so that $g^{-1}hg \in G_{(\Delta)}$. From this it follows that $g^{-1}G_{(\Gamma)}g \subseteq G_{(\Delta)}$. The reverse inclusion follows from a similar argument. $\square$

Some more examples:

3(f) The symmetric group S_n is clearly n-transitive, since it contains *all* the permutations on a set of n elements.

3(g) The alternating group A_n, however, is only $(n-2)$-transitive. To see this, let $\{\alpha_1, \alpha_2, \ldots, \alpha_{n-2}\}$ and $\{\beta_1, \beta_2, \ldots, \beta_{n-2}\}$ be two sets of distinct points in Ω. Let α_{n-1} and α_n denote the two elements of Ω not in the set $\{\alpha_i \mid 1 \le i \le n-2\}$. Define β_{n-1} and β_n similarly. Then since S_n is n-transitive, we can find $g, h \in S_n$ such that $\alpha_i^g = \alpha_i^h = \beta_i$ for $1 \le i \le n-2$, and such that $\alpha_{n-1}^g = \alpha_n^h = \beta_{n-1}$ and $\alpha_n^g = \alpha_{n-1}^h = \beta_n$. Then clearly, $g = h(\beta_{n-1}\beta_n)$. From this it follows that one of g or h must be even and this even permutation takes α_i to β_i for $1 \le i \le n-2$.

3(h) For $n \ge 3$, consider the action of S_n on the set $\Omega^{\{2\}}$ of unordered pairs from the set $\{1, 2, \ldots, n\}$ given by $\{a, b\}^g = \{a^g, b^g\}$. This action is transitive, as S_n is 2-transitive. The point stabiliser is easily seen to be the group $C_2 \times S_{n-2}$.

3(i) We have already seen that the group $\mathrm{GL}(2, \mathbb{R})$ is transitive in its action on $\mathbb{R}^2 \setminus \{(0,0)\}$ as defined earlier. It is however, not 2-transitive because, for example, we cannot find a matrix in $\mathrm{GL}(2, \mathbb{R})$ that maps the non-zero vector $(x, 0)$ to itself while mapping the vector $(y, 0)$ (with $y \ne x, 0$) to $(0, y)$.

A group G acting on a set Ω is said to be *multiply transitive* if it is k-transitive for some $k \ge 2$. We end this section by strengthening the notion of multiple transitivity.

Definition 3.15 A k-transitive group G is said to be *sharply k-transitive* on Ω if for any two k-tuples of distinct elements from Ω there is exactly one element $g \in G$ which takes the first into the second. Equivalently, a k-transitive G is *sharply k-transitive* on Ω if the pointwise stabiliser of every k-set from Ω is the identity subgroup.

Note that a sharply 1-transitive group is the same as a regular group. Tits (1952) showed that there are no infinite sharply k-transitive groups for $k \ge 4$.

3.4   Homogeneity

A weakening of the notion of transitivity leads us to the notion of homogeneity in permutation groups.

Definition 3.16 A G-space Ω is said to be *k-homogeneous* if for every $\Gamma, \Delta \subseteq \Omega$ with $|\Gamma| = |\Delta| = k$ there is some $g \in G$ such that $\Gamma^g = \Delta$.

An infinite space is said to be *highly homogeneous* if it is k-homogeneous for every $k \in \mathbf{N}$.

Note that Corollary 3.14 also holds under the weaker assumption that Ω is k-homogeneous. Clearly, a k-transitive space is always k-homogeneous but the converse is not always true as the following example will show.

Another example:

3(j) Consider the group $G := \mathrm{Aut}\,(\mathbb{Q}, <)$ of order automorphisms of the rationals, namely,

$$\{g \in \mathrm{Sym}\,(\mathbb{Q}) \mid p < q \text{ if and only if } p^g < q^g, \text{ for all } p, q \in \mathbb{Q}\}.$$

This group is k-homogeneous for every $k \in \mathbf{N}$. To see this, let Γ and Δ be finite subsets of $\mathbb{Q}$ of same size n. We will explicitly define a piecewise linear automorphism g which maps Γ onto Δ. Let

$$\Gamma = \{x_1, \ldots, x_n\} \text{ where } x_1 < x_2 < \ldots < x_n,$$

and

$$\Delta = \{y_1, \ldots, y_n\} \text{ where } y_1 < y_2 < \ldots < y_n.$$

Define intervals $A_i := (x_i, x_{i+1})$ and $B_i := (y_i, y_{i+1})$ for each $i = 1, 2, \ldots, n - 1$, and set $A_0 := (-\infty, x_1)$, $A_n := (x_n, \infty)$, $B_0 := (-\infty, y_1)$, $B_n := (y_n, \infty)$. Define $f_i : A_i \to B_i$, for $i = 0, 1, \ldots, n$ as follows:

$$
\begin{aligned}
f_0 : x &\;\mapsto\; y_1 + (x - x_1) \\
f_i : x &\;\mapsto\; y_i + (x - x_i)(y_{i+1} - y_i)/(x_{i+1} - x_i) \\
f_n : x &\;\mapsto\; y_n + (x - x_n)
\end{aligned}
$$

Define $g : \mathbb{Q} \to \mathbb{Q}$ by

$$x_i^g = y_i, \text{ for } i = 1, \ldots, n \text{ and}$$
$$g|_{A_i} = f_i, \text{ for } i = 0, \ldots, n.$$

Clearly $g \in \text{Aut}(\mathbb{Q}, <)$ and $\Gamma^g = \Delta$.

But this group is not even 2-transitive, as given a pair of rational numbers p, q with $p < q$ no order automorphism of $\mathbb{Q}$ can map p to q while mapping q back to p.

We leave the proof of the following theorem as an exercise, as it is similar to the proof of Theorem 3.13.

Theorem 3.17 *Let Ω be a transitive G-space and $\alpha \in \Omega$. Then G is $(k+1)$-homogeneous on Ω if G_α is k-homogeneous on $\Omega \setminus \{\alpha\}$.* $\square$

Note however, that the converse of the last theorem is not necessarily true, as can be seen by considering the group $\text{Aut}(\mathbb{Q}, <)$. We will use the following number-theoretic lemma in the proof of Theorem 3.19.

Lemma 3.18 *Let p be a prime number, r be a natural number and $\alpha \geq 1$ be a natural number. Then p does not divide $\binom{p^\alpha r - 1}{p^\alpha - 1}$.*

Proof: We have

$$
\begin{aligned}
\binom{p^\alpha r - 1}{p^\alpha - 1} &= \frac{(p^\alpha r - 1)!}{(p^\alpha (r - 1))! \, (p^\alpha - 1)!} \\
&= \frac{(p^\alpha r - 1)(p^\alpha r - 2) \cdots (p^\alpha r - (p^\alpha - 1))}{(p^\alpha - 1)!} \\
&= \frac{\prod_{i=1}^{p^\alpha - 1}(p^\alpha r - i)}{(p^\alpha - 1)!} \\
&= \frac{\prod_{i=1}^{p^\alpha - 1}(p^\alpha r - i)}{\prod_{i=1}^{p^\alpha - 1} i} \\
&= \prod_{i=1}^{p^\alpha - 1} \frac{p^\alpha r - i}{i}.
\end{aligned}
$$

Looking at one of the factors in the last expression we see that p^β divides the numerator, $p^\alpha r - i$, if and only if p^β divides the denominator, i. Thus if β is chosen so that p^β is the highest power of p that divides $\prod_{i=1}^{p^\alpha-1}(p^\alpha r - i)$ then p^β is also the highest power of p that divides $\prod_{i=1}^{p^\alpha-1} i = (p^\alpha - 1)!$. Thus the fraction $\dfrac{\prod_{i=1}^{p^\alpha-1} p^\alpha r - i}{\prod_{i=1}^{p^\alpha-1} i}$ is not divisible by p. □

The following theorem can be traced back to Brown (1959). The proof we give is from Wielandt (1967). A different proof, and a generalization, can be found in Cameron (1976).

Theorem 3.19 *If $m \leq k$ and $2k \leq |\Omega|$, then every k-homogeneous group on Ω is also m-homogeneous.*

Proof: Suppose G is k-homogeneous on Ω. We prove that G is then also $(k-1)$-homogeneous, and it will follow, by induction, that G is m-homogeneous.

Set $n := |\Omega|$ and for i with $1 \leq i \leq n$ set

$$\Omega^{\{i\}} := \{\Gamma \subseteq \Omega \mid |\Gamma| = i\}.$$

Let Θ denote an orbit of G on $\Omega^{\{k-1\}}$. A subset Γ of Ω of size k contains k subsets of size $k-1$, say $\Gamma_1, \ldots, \Gamma_k$. Suppose that k' of these, $\Gamma_1, \ldots, \Gamma_{k'}$, are contained in Θ. If Δ is some other set from $\Omega^{\{k\}}$, then there is some element $g \in G$ such that $\Gamma^g = \Delta$. Now it is evident that also Δ contains precisely k' subsets of size $k-1$ that are in Θ, namely $\Gamma_1^g, \ldots, \Gamma_{k'}^g$.

To prove our theorem it is enough to show that $k' = k$ because then all the elements in $\Omega^{\{k-1\}}$ will be in Θ. This will be done by showing that k divides k'.

Choose some natural number s, such that $k \leq s \leq n$ and let $\Sigma \in \Omega^{\{s\}}$. Let us count all pairs (Γ, Δ) such that $\Gamma \subset \Delta \subseteq \Sigma$, the number of elements in Δ is k and $\Gamma \in \Theta$. The number of choices for Δ is clearly $\binom{s}{k}$ and when Δ has been fixed, there are k' different choices for Γ. Thus the number of all such pairs is $\binom{s}{k}k'$. If we instead start by choosing Γ, then we have $s - k + 1$ different choices for Δ. (We get Δ by adding some element in $\Sigma \setminus \Gamma$ to Γ.) Therefore $s - k + 1$ divides $\binom{s}{k}k'$. Now let p be a prime and suppose p^α divides k, for some $\alpha \geq 1$.

Set $s := k + p^\alpha - 1$. Since $2k \leq n$ we know that $s < 2k \leq n$. We have shown that $s - k + 1 = p^\alpha$ divides

$$\binom{s}{k} k' = \binom{k + p^\alpha - 1}{p^\alpha - 1} k'.$$

But now we can apply Lemma 3.18. We set $r = \frac{k}{p^\alpha} + 1$. Then $p^\alpha r - 1 = k + p^\alpha - 1 = s$, and by the lemma p does not divide $\binom{p^\alpha r - 1}{p^\alpha - 1} = \binom{s}{k}$, so p^α divides k'. Whence k divides k' and we must have $k' = k$. $\square$

EXERCISES:

3(i) Let $\theta : \Omega \to \Omega'$ be a G-morphism. Show that if Ω_i is an orbit in Ω then $\Omega_i \theta$ is an orbit in Ω'.

3(ii) Consider the action of S_n on $\Omega^{\{2\}}$ described in Example 3(h). Prove that the action is not 2-homogeneous. Furthermore, show that the action has 2 orbits on unordered 2-sets from $\Omega^{\{2\}}$.

3(iii) Let $G := S_n$ act naturally on the set $\Omega := \{1, \ldots, n\}$. This gives us also an action of G on the set of k-element subsets of Ω. Let $\Delta \subseteq \Omega$ with $|\Delta| = k$. Describe the orbits of $G_{\{\Delta\}}$ on the set of k-element subsets of Ω and count the number of elements in each orbit.

3(iv) Let Ω be a countably infinite set. Show that $\text{Sym}(\Omega)$ has countably many orbits on the power set $\wp(\Omega)$ and describe them.

3(v) Let Ω be a transitive G-space. Suppose that $H \leq G$ with $|G : H| = k$. Show that if H has t orbits on Ω then $t \leq k$.

3(vi) Let G be a group acting faithfully and transitively on a finite set Ω. Show that if α is some element of Ω then $|G| = |\Omega| \cdot |G_\alpha|$. Show also that the action is regular if and only if $|G| = |\Omega|$.

3(vii) Let G be a group acting faithfully on a finite set Ω. For $g \in G$ define $\text{fix}_\Omega(g)$ as the number of points $\alpha \in \Omega$ that are fixed by g, i.e. such that $\alpha^g = \alpha$.

Show that if G is transitive then

$$\frac{1}{|G|} \sum_{g \in G} \text{fix}_\Omega(g) = 1.$$

[Hint: How many permutations fix a given point α in Ω?] This simple result is very important both in combinatorics and group theory. It is often referred to as "Burnside's Lemma" but the attribution is mistaken. Those who are interested in learning how mistaken attributions get into mathematical literature can consult Neumann (1979).

3(viii) Generalise the result in the previous exercise to show that

$$\frac{1}{|G|} \sum_{g \in G} \text{fix}_\Omega(g)$$

is equal to the number of orbits of G on Ω.

3(ix) Suppose that G acts transitively on a finite set Ω and $|\Omega| > 1$. Show that G must always contain a permutation g that has no fixed points, i.e. $\text{fix}_\Omega(g) = 0$.

3(x) Show that S_n is sharply n-transitive and that A_n is sharply $(n - 2)$-transitive.

Chapter 4

Primitivity

We have already seen how a group action on a set can be broken up into transitive actions. We can also try to break a transitive action up into a form of a 'product' of transitive actions. We can continue to do so till we arrive at actions which cannot be further broken up (of course, sometimes the process does not terminate after finitely many steps). These are the primitive actions which we will consider in this chapter. As in Chapter 3, we refer to Wielandt (1960, 1964) for more information on this subject.

4.1 G-congruences

Definition 4.1 Let Ω be a transitive G-space and let $\approx$ be an equivalence relation on Ω. If $\alpha \approx \beta \Leftrightarrow \alpha^g \approx \beta^g$ for all $\alpha, \beta \in \Omega$ and for all $g \in G$, we say that $\approx$ is a G-congruence on Ω.

In this case we can also say that the relation $\approx$ is G-invariant. The equivalence classes are called the $\approx$-classes (or simply classes) of the G-congruence $\approx$.

A G-congruence is said to be non-trivial if there is a class with more than one element and it is said to be proper if there is more than one class.

Examples :

4(a) The relation $\alpha \approx \beta \Leftrightarrow \alpha = \beta$ is always a G-congruence. This congruence is called the *trivial congruence*. Here the $\approx$-classes are singleton sets.

4(b) The relation $\alpha \approx \beta$ for all $\alpha, \beta \in \Omega$ is also always a G-congruence. This congruence is known as the *universal congruence* (or sometimes an *improper congruence*) as there is only one $\approx$-class in this case consisting of the whole set Ω.

4(c) For $G := \mathrm{GL}(2, \mathbb{R})$ acting on $\mathbb{R}^2 \backslash \{(0,0)\}$, the relation $v_1 \approx v_2$ if and only if v_1 and v_2 are linearly dependent is a G-congruence. If we identify $\mathbb{R}^2$ with the plane, then the $\approx$-classes are lines passing through (but not containing) the origin.

The relation between normal subgroups of a transitive group G and G-congruences is given by the following theorem.

Theorem 4.2 *Let Ω be a transitive G-space and let N be a normal subgroup of G. Then the orbits of N are the $\approx$-classes of a G-congruence $\approx$ on Ω.*

Proof: Let us define $\alpha \approx \beta$ if and only if there is some $x \in N$ such that $\alpha^x = \beta$. It is clear then that $\approx$ is an equivalence relation and the equivalence classes are just the orbits of N. Suppose that $\alpha \approx \beta$ and let $x \in N$ be such that $\alpha^x = \beta$. We want to show that $\alpha^g \approx \beta^g$ for $g \in G$. Now $\alpha^{xg} = \beta^g$. Because N is normal in G there is some $x' \in N$ such that $xg = gx'$. Then $\beta^g = \alpha^{xg} = \alpha^{gx'}$ which implies that $\alpha^g \approx \beta^g$ as $x' \in N$.

Conversely, if $\alpha^g \approx \beta^g$ then $\alpha^{gx} = \beta^g$ for some $x \in N$ and so $\alpha^{gxg^{-1}} = \beta$ which implies that $\alpha \approx \beta$ as $gxg^{-1} \in N$. $\square$

If ρ is a G-congruence on a set Ω, we say that α and β are *ρ-equivalent*, and write $\alpha \equiv \beta \,(\mathrm{mod}\,\rho)$, when α and β belong to the same ρ-class, and write $\alpha \not\equiv \beta \,(\mathrm{mod}\,\rho)$ when they don't. We also sometimes denote the ρ-class containing an element θ of Ω by $\rho(\theta)$. That is, $\rho(\theta) := \{\omega \mid \omega \equiv \theta \,(\mathrm{mod}\,\rho)\}$. Note that in a transitive G-space Ω, all classes will have the same number of elements. If Ω is finite, since the classes fill up the whole space, the size of any class must be a divisor of $|\Omega|$.

4.2 Primitive spaces

Definition 4.3 Let Ω be a transitive G-space. If there are no non-trivial, proper G-congruences then Ω is said to be a *primitive G-space* (or the action of G on Ω is said to be *primitive*). Otherwise, G is said to act *imprimitively* on Ω.

The above definition implies that if $\approx$ is a G-congruence on a primitive G-space Ω then either $\approx$ is trivial (and all $\approx$-classes have just one element) or it is universal (and there is only one $\approx$-class, the whole of Ω).

Examples :

4(d) Let C_p be the cyclic group with p elements, for some prime p. Then C_p is primitive in its action on itself via the right regular representation, since p has no non-trivial proper divisors. More generally, if Ω is a transitive G-space with $|\Omega| = p$, where p is a prime, then it is primitive.

4(e) Any 2-transitive permutation group is primitive. Suppose that $\approx$ is a non-trivial G-congruence, and α and β are distinct points in Ω with $\alpha \approx \beta$. Then for any other pair of distinct points $\gamma, \delta \in \Omega$ we can find $g \in G$ such that $\alpha^g = \gamma$ and $\beta^g = \delta$. But then $\gamma \approx \delta$, and so $\approx$ is the universal relation.

The same argument can be used to show that a 2-homogeneous group (of degree > 2) is primitive. As particular cases of these results, we have that S_n is primitive for $n \geq 2$ and A_n is primitive for $n \geq 3$.

We now prove a theorem which relates primitivity of a group to transitivity of its normal subgroups.

Theorem 4.4 *Let G be a group acting faithfully and primitively on a set Ω. If N is any non-trivial normal subgroup of G then N is transitive.*

Recall that the action of G is faithful if every non-identity element of G moves at least one point of Ω.

Proof: By Theorem 4.2, the orbits of N define a G-congruence on Ω. Since G is primitive, this G-congruence must be either trivial or universal. If it is trivial, then $\alpha^N = \alpha$ for all $\alpha \in \Omega$ which contradicts the assumption that the action is faithful. So it must be the universal congruence, in which case $\alpha^N = \Omega$ for any $\alpha \in \Omega$ which means that N is transitive on Ω. $\square$

From now on we shall assume that G acts transitively on Ω.

Definition 4.5 A subset $\Delta \subseteq \Omega$ is called a *block* if for every $g \in G$ either $\Delta \cap \Delta^g = \emptyset$ or $\Delta = \Delta^g$. A block is said to be *non-trivial* if $|\Delta| > 1$ and *proper* if $\Delta \neq \Omega$.

We also have the following definition:

Definition 4.6 A subset $\Delta \subseteq \Omega$ is said to *separate points* if for every pair α, β of distinct points in Ω there is some $g \in G$ such that Δ^g contains one of α and β but not both.

Clearly singleton sets will always separate points.

Theorem 4.7 *For a G-space Ω with $|\Omega| > 1$, the following are equivalent:*

  (i) *Ω is primitive.*

 (ii) *Every non-empty proper subset of Ω separates points.*

(iii) *Ω has no non-trivial proper blocks.*

(iv) *For every $\alpha \in \Omega$, the subgroup G_α is a maximal subgroup of G.*

Proof:
(i)$\Rightarrow$(ii)
Let Δ be a non-empty proper subset of Ω. For $\alpha, \beta \in \Omega$ we set $\alpha \approx \beta$ if for every $g \in G, \alpha \in \Delta^g$ if and only if $\beta \in \Delta^g$. It is easy to prove that then $\approx$ is an equivalence relation. To show that it is a G-congruence, let $\alpha \approx \beta$ and $g \in G$. We want to show that $\alpha^g \approx \beta^g$. So suppose $\alpha^g \in \Delta^h$ for some $h \in G$. Then $\alpha \in \Delta^{hg^{-1}}$. Since $\alpha \approx \beta$ we must have $\beta \in \Delta^{hg^{-1}}$, so that $\beta^g \in \Delta^h$. Therefore $\approx$ is a G-congruence.

But since Ω is primitive, $\approx$ is either trivial or universal. The latter possibility implies that $\Delta = \Omega$, contrary to assumption. Therefore $\approx$

must be the trivial congruence, and so Δ separates points.

(ii)$\Rightarrow$(iii)

Suppose Δ is a non-trivial block and let α, β be two distinct points of Δ. Then if $\alpha \in \Delta^g$ we have $\Delta \cap \Delta^g \neq \emptyset$. Since Δ is a block, this implies $\Delta = \Delta^g$. Therefore, whenever $\alpha \in \Delta^g$, we must also have $\beta \in \Delta^g$. Since by (ii), every non-empty proper subset of Ω separates points, we must have $\Delta = \Omega$.

(iii)$\Rightarrow$(iv)

Suppose $G_\alpha \leq H \leq G$ and set $\Delta := \alpha^H$, the orbit of α under H. We will show that Δ is a block. Suppose $\beta \in \Delta \cap \Delta^g$ for some $g \in G$. Then $\beta \in \alpha^H$ and $\beta \in \alpha^{Hg}$, so that there exist elements $h, h' \in H$ such that $\beta = \alpha^h = \alpha^{h'g}$. But then $\alpha = \alpha^{h'gh^{-1}}$ so that $h'gh^{-1} \in G_\alpha \leq H$. This implies $g \in H$ and so $\Delta^g = \Delta$. This proves that Δ is a block. By (iii) this means that either $\Delta = \{\alpha\}$ or $\Delta = \Omega$. In the first case, $H = G_\alpha$ (as H must stabilise α and $G_\alpha \leq H$). In the second, we have $\alpha^H = \Omega$ which means that H is transitive on Ω. In this case, for $\beta \in \Omega$ let $h_\beta \in H$ be such that $\alpha^{h_\beta} = \beta$. Then clearly,

$$G = \bigcup_{\beta \in \Omega} G_\alpha h_\beta \leq H,$$

and so $H = G$.

(iv)$\Rightarrow$(i)

We start with a G-congruence $\approx$. Our aim is to prove that $\approx$ is either trivial or universal. Let Δ be the $\approx$-class containing α and set

$$H := G_{\{\Delta\}} = \{g \in G \mid \Delta^g = \Delta\}.$$

Now $G_\alpha \leq H$. This is because if $g \in G_\alpha$ and $\delta \in \Delta$ then $\delta \approx \alpha$ and so $\alpha = \alpha^g \approx \delta^g$. The assumption that G_α is a maximal subgroup of G now tells us that either $H = G_\alpha$ or $H = G$. In the first case $\Delta = \{\alpha\}$ and the G-congruence $\approx$ is trivial, and in the second case $\Delta = \Omega$ and the G-congruence $\approx$ is universal. $\square$

It is possible to carry further the analysis of the connections between G-congruences and subgroups of G. Suppose that ρ and σ are two G-congruences on Ω. Define $\rho \leq \sigma$ if for every $\omega \in \Omega$ the ρ-class $\rho(\omega)$ is contained in the σ-class $\sigma(\omega)$.

Theorem 4.8 *Let Ω be a transitive G-space, and let $\alpha \in \Omega$. There exists a one-one correspondence between the set of G-congruences and the set of subgroups H of G such that $G_\alpha \le H$.*

Furthermore, if ρ, σ are two G-congruences with $\rho \le \sigma$ and $H(\rho)$, $H(\sigma)$ are the corresponding subgroups of G, then $H(\rho) \le H(\sigma)$. The subgroup corresponding to the trivial congruence is G_α and the subgroup corresponding to the universal congruence is G.

Proof: Let ρ be a G-congruence on Ω. Define

$$H(\rho) := \{g \in G | \alpha^g \equiv \alpha \,(\mathrm{mod}\,\rho)\}.$$

From the definition it is clear that $G_\alpha \le H(\rho)$. Now let $\beta \in \rho(\alpha)$ and $h \in H(\rho)$. Then

$$\beta^h \equiv \alpha^h \equiv \alpha \,(\mathrm{mod}\,\rho).$$

So β is also in $\rho(\alpha)$. We see that $h \in H(\rho)$ precisely if $\rho(\alpha)^h = \rho(\alpha)$. Looking at this makes it clear that $H(\rho)$ is a subgroup of G.

We still have to check that this gives us a one-one correspondence. Suppose ρ and σ are two G-congruences so that $H(\rho) = H(\sigma)$. The definition of $H(\rho)$ says that $\rho(\alpha) = \alpha^{H(\rho)}$, and in the same way we have $\sigma(\alpha) = \alpha^{H(\sigma)}$, so that $\rho(\alpha) = \sigma(\alpha)$. Let $\beta \in \Omega$ and take $g \in G$ such that $\alpha^g = \beta$. Then, using the properties of G-congruences,

$$\rho(\beta) = \rho(\alpha^g) = \rho(\alpha)^g = \sigma(\alpha)^g = \sigma(\alpha^g) = \sigma(\beta).$$

Hence $\rho = \sigma$. The last bit here is to show that if H is a subgroup of G that contains G_α then there is a G-congruence ρ such that $H = H(\rho)$. We define ρ such that $\rho(\alpha) := \alpha^H$, and if $\beta \in \Omega$ and $g \in G$ such that $\alpha^g = \beta$ then we set $\rho(\beta) := \rho(\alpha)^g$. Here we have to show that the choice of g is immaterial, and that ρ is indeed a G-congruence. Suppose $g, g' \in G$ with $\beta = \alpha^g = \alpha^{g'}$. Clearly $\alpha^{g'g^{-1}} = \beta^{g^{-1}} = \alpha$, therefore $g'g^{-1} \in G_\alpha \le H$. Hence $\rho(\alpha) = \rho(\alpha)^{g'g^{-1}}$ which implies $\rho(\alpha)^g = \rho(\alpha)^{g'}$. So ρ is well defined. To show that ρ is a G-congruence we need to show that $\rho(\omega)^g = \rho(\omega^g)$ for every $g \in G$ and every $\omega \in \Omega$. Write $\omega = \alpha^h$ for some $h \in G$. Then

$$\rho(\omega)^g = \rho(\alpha^h)^g = (\rho(\alpha)^h)^g = \rho(\alpha)^{hg} = \rho(\alpha^{hg}) = \rho(\omega^g).$$

Finally we show that if $\rho \le \sigma$ then $H(\rho) \le H(\sigma)$. But $\rho \le \sigma$ means precisely that if $\alpha \equiv \beta \,(\mathrm{mod}\,\rho)$ then $\alpha \equiv \beta \,(\mathrm{mod}\,\sigma)$. Looking again at the definition for $H(\rho)$ it is clear that $H(\rho) \le H(\sigma)$.

The last statement in the theorem is also clear from the definition of $H(\rho)$. □

Note that Theorem 4.7 (iv) is just a special case of the last theorem. Using the language of lattices, we now state a slightly stronger result as an exercise. A *lattice* is defined to be a partially ordered set in which every pair of elements has a least upper bound and a greatest lower bound.

EXERCISE:

4(i) If G acts on the right cosets of a subgroup H of G by right multiplication, then show that the lattice of G-congruences is isomorphic to the lattice of intermediate subgroups (that is, subgroups K such that $H \leq K \leq G$).

4.3 Extensions of primitivity

We have already extended the notion of transitivity to define k-transitivity in Section 3.3. We end this chapter by similarly extending the notion of primitivity.

Definition 4.9 Let $k \in \mathbf{N}$. A group G acting on a set Ω is said to be k-*primitive* if it is k-transitive on Ω, and for all distinct points $\alpha_1, \alpha_2, \ldots, \alpha_{k-1} \in \Omega$ their pointwise stabiliser $G_{\alpha_1,\alpha_2,\ldots,\alpha_{k-1}}$ is primitive on the set

$$\Omega \setminus \{\alpha_1, \alpha_2, \ldots, \alpha_{k-1}\}.$$

The proof of the following lemma is immediate from the fact that a 2-transitive group is primitive.

Lemma 4.10 *A k-transitive group is at least $(k-1)$-primitive.* □

The above lemma also implies that if a group is not k-primitive, it cannot be $(k+1)$-transitive. We shall use these facts repeatedly in the later chapters.

EXERCISES:

4(ii) Consider the action of S_n on $\Omega^{\{2\}}$ described in Example 3(h). Is the action primitive?

4(iii) As a variant to the last exercise, consider the natural action of S_n on $\Omega^{(2)}$ (or simply Ω^2), the set of *ordered* pairs from $\{1, 2, \ldots, n\}$. Show that this action is imprimitive, and that the relations ρ_i defined on $\Omega^{(2)}$ by $(a_1, a_2) \equiv (b_1, b_2) \pmod{\rho_i}$ if and only if $a_i = b_i$, for $i = 1, 2$ are congruences.

4(iv) The *infinite dihedral group* D_∞ can be defined as the group of all permutations g of $\mathbf{Z}$ such that $|i - j| = |i^g - j^g|$ for all $i, j \in \mathbf{Z}$. Is the action of D_∞ on $\mathbf{Z}$ primitive?

4(v) When is the regular action of a group G on itself primitive?

4(vi) Let $\Omega = \mathbf{R}^n$ and let G be the the group of all permutations of Ω such that $\|x - y\| = \|x^g - y^g\|$ for all $x, y \in \mathbf{R}$. Such a permutation is called an *isometry* and G is the group of all isometries. Is this action primitive?

4(vii) A permutation group is said to be of *degree* n if it is a subgroup of the symmetric group of degree n. List all primitive permutation groups of degree less than or equal to 5.

4(viii) Prove that if $H \leq G$ then $\cos(G : H)$ is a primitive G-space if and only if H is a maximal subgroup of G.

4(ix) Suppose that G acts primitively on Ω, and $\alpha, \beta \in \Omega$. Show that then either G is a cyclic group of prime order acting regularly on Ω or $G_\alpha \neq G_\beta$. Also show that in the second case we have $G = \langle G_\alpha \cup G_\beta \rangle$.

4(x) Suppose that G acts faithfully and primitively on Ω, and that G has a non-trivial abelian normal subgroup A. Prove that then the following holds.
   (a) A acts regularly on Ω.
   (b) If $\alpha \in \Omega$ then $G = G_\alpha A$ and $G_\alpha \cap A = \{1\}$.
   (c) $A = C_G(A)$ where $C_G(A)$ is the *centraliser* of A in G and is defined to be the set $\{g \in G \mid ga = ag \text{ for all } a \in A\}$.
   (d) If N is a non-trivial normal subgroup of G then $A \leq N$.

Chapter 5

Suborbits and Orbitals

We have already seen the concept of an orbit in Chapter 3. We now describe the related notions of suborbits and orbitals of a permutation group G acting on a set Ω.

5.1 Suborbits and orbitals

Definition 5.1 Let Ω be a transitive G-space.

(i) Let $\alpha \in \Omega$. Then the stabiliser G_α also acts on Ω. The orbits of G_α are called the *suborbits* (or the (α)-*suborbits*) of G.

(ii) The group G also acts on the set $\Omega^2 = \Omega \times \Omega$ via the action given by $(\omega_1, \omega_2)^g = (\omega_1^g, \omega_2^g)$ (cf. Ex. 4(iii)). The orbits of G on Ω^2 are known as the *orbitals* (or the G-*orbitals*) of G.

Note that $\{\alpha\}$ is always a G_α-orbit and is called the *trivial suborbit* of G. The subset $\Delta_0 := \{(\omega, \omega) \mid \omega \in \Omega\}$ of Ω^2 is always an orbital and is called the *diagonal orbital* of G.

Examples :

5(a) Let $G = \mathrm{Sym}(\Omega)$ with $|\Omega| > 1$. Then for any $\alpha \in \Omega$ the suborbits are $\{\alpha\}$ and $\Omega \setminus \{\alpha\}$. This is because $\mathrm{Sym}(\Omega)$ is 2-transitive on Ω so that G_α is transitive on $\Omega \setminus \{\alpha\}$. Using the same reasoning it is easy to see that there are just two orbitals of $\mathrm{Sym}(\Omega)$ on Ω^2, namely, Δ_0 (as defined above) and $\Omega^2 \setminus \Delta_0$. Clearly, the same will hold for any 2-transitive group.

39

5(b) At the other extreme, let G act regularly on Ω. Then G is transitive and $G_\alpha = \{1\}$ for any $\alpha \in \Omega$. Therefore the suborbits of G are clearly the singleton subsets $\{\omega\}$ of Ω. There are $|\Omega|$ of them. There are as many orbitals, each containing one and only one element from the set $\{(\alpha, \omega) \mid \omega \in \Omega\}$.

5(c) Let $G := \operatorname{Aut}(\mathbb{Q}, <)$ and let $p \in \mathbb{Q}$. We have already seen that G is highly homogeneous (cf. Eg. 3(j)). The orbits of G_p are the sets $\{p\}$, $\{q \in \mathbb{Q} \mid q < p\}$ and $\{q \in \mathbb{Q} \mid q > p\}$. The corresponding orbitals are $\{(q, q)\}$, $\{(q_1, q_2) \mid q_1 < q_2\}$ and $\{(q_1, q_2) \mid q_1 > q_2\}$.

The fact that the numbers of suborbits and orbitals in the above examples are the same is no coincidence, but is a consequence of the following general theorem.

Theorem 5.2 *Let Ω be a transitive G-space and let $\alpha \in \Omega$. Then there is a natural one-to-one correspondence between the orbits of G_α on Ω and the G-orbits on Ω^2. The trivial suborbit corresponds to the diagonal orbital.*

Sketch Proof: Given a subset Δ of Ω^2 define

$$\Delta(\alpha) := \{\beta \in \Omega \mid (\alpha, \beta) \in \Delta\}.$$

Let Δ be an orbital and let $\beta_1, \beta_2 \in \Delta(\alpha)$. Then there exists some $g \in G$ which takes (α, β_1) to (α, β_2). The element g therefore fixes α (and hence belongs to G_α) and takes β_1 to β_2. Clearly $\Delta(\alpha) \neq \emptyset$ because G is transitive. Thus we conclude that $\Delta(\alpha)$ is contained in a suborbit of G. It is easy to see that the $\Delta(\alpha)$ is in fact a suborbit of G.

The mapping $\Delta \mapsto \Delta(\alpha)$ is the required correspondence. The map is well-defined and injective as orbits are disjoint. It is surjective because the union of the G-orbitals is the whole of Ω^2, so that the union of the corresponding suborbits is the whole of Ω. Also the trivial suborbit corresponds to the diagonal orbital under this map. This is all we need to complete the proof. $\square$

We end this section with a lemma that illustrates the use of the concepts of suborbits and orbitals in permutation group theory.

Lemma 5.3 *Suppose that G is primitive on Ω and has some nontrivial finite suborbits, but only finitely many. Then Ω is finite.*

Proof: Let $\Gamma_0 := \{\alpha\}, \Gamma_1, \ldots, \Gamma_s$ be the finite G_α-orbits and let Δ_i be the corresponding orbitals (under the correspondence defined in the last theorem), for $0 \le i \le s$. Define a binary relation $\approx$ on Ω by declaring that $\beta \approx \gamma$ if $(\beta, \gamma) \in \Delta_i$ for some $i = 0, 1, \ldots, s$. Also, for each $\theta \in \Omega$ put $\Gamma(\theta) = \{\delta \mid \theta \approx \delta\}$. We claim that $\approx$ is a G-congruence.

It is certainly G-invariant since $(\beta, \gamma) \in \Delta_i$ implies $(\beta, \gamma)^g \in \Delta_i$ for all $g \in G$. Since Δ_0 contains all pairs (β, β), the relation $\approx$ is also reflexive. To prove transitivity of the relation $\approx$, suppose $\alpha \approx \beta$ and $\beta \approx \gamma$. Then there are only finitely many possibilities for β^g with $g \in G_\alpha$ and there are only finitely many for γ. Hence the G_α-orbit containing γ is finite, which means that $\alpha \approx \gamma$. Finally to show symmetry, let $\beta \approx \gamma$. Then by transitivity of the relation $\approx$, we have $\Gamma(\gamma) \subseteq \Gamma(\beta)$. But by transitivity of G, it follows that these are finite sets of the same size. So $\Gamma(\gamma) = \Gamma(\beta)$. Since $\beta \in \Gamma(\beta)$ so $\beta \in \Gamma(\gamma)$ and therefore $\gamma \approx \beta$.

Since G is primitive, this completes the proof of the theorem, as it follows that $\approx$ is universal, which in turn implies that $\Omega = \bigcup_{i=0}^{s} \Gamma_i$ is finite. $\square$

5.2 Orbital graphs and primitivity

Ideas from graph theory have powerful applications to group actions. Let us first recall the definition of a graph. A *graph* Γ is a pair (V, E) where V is a non-empty set and E is a set of pairs from V. The elements of V are called the *vertices* while those in E are called the *edges* of the graph Γ. A graph is *directed* if the pairs in E are ordered. The edges of a directed graph are also called *directed edges*. Two vertices are said to be *adjacent* if they are joined by an edge. Given a vertex, all the vertices adjacent to it are called its *neighbours*. A (directed) edge joining a vertex to itself is called a *loop*. We can now make the following definition.

Definition 5.4 Let G be a group acting on a set Ω and let Δ be an orbital of G. We define the *orbital graph* of G with respect to the orbital Δ to be the directed graph that has Ω as its vertex set and Δ as its set of edges.

If $(\alpha, \beta) \in \Delta$ then the orbital of G containing (α, β) is the set $(\alpha, \beta)^G$. Also note that if Δ_0 is the diagonal orbital then its orbital graph has a loop at each vertex and no other edges. We call this the *trivial* orbital graph. All others will be called *non-trivial*. Finally, we mention that in a directed graph, edges are drawn as arrows. So if $(\alpha, \beta) \in \Delta$ then we have an arrow going from the vertex α to β in the orbital graph of Δ.

Examples :

5(d) If $G := C_5 = \langle (1\ 2\ 3\ 4\ 5) \rangle$, the cyclic group of order 5, and $\Delta := (1, 2)^G$ then the orbital graph of Δ is a graph with vertices $1, 2, 3, 4$ and 5 and edges from vertex i to $i + 1$ identifying 6 and 1. It looks simply like an oriented pentagon (see Fig. 1).

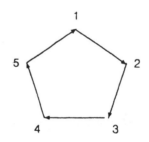

Figure 1

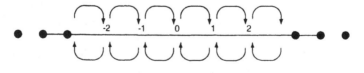

Figure 2

5(e) Consider the infinite dihedral group D_∞ acting on $\mathbf{Z}$ (compare Ex. 4(iv)). It is generated by the two elements

$$A := (\ldots -2 -1\,0\,1\,2 \ldots) \text{ and } B := (0)(-1\,1)(-2\,2)\ldots.$$

Set $\Delta := (0,1)^{D_\infty}$. Then the orbital graph looks like a copy of $\mathbf{Z}$ with two edges between consecutive points i and $(i+1)$, one going from i to $(i+1)$ and the other in the opposite direction (see Fig. 2). This is because B gives the edge $(0,-1)$ from the edge $(0,1)$; then A^i yields $(i,i+1)$ from $(0,1)$, and A^{i+1} yields $(i+1,i)$ from $(0,-1)$.

Definition 5.5 A *directed path* from a vertex α to β in a graph Γ is a sequence

$$\alpha = \alpha_0, \alpha_1, \ldots, \alpha_n = \beta$$

of vertices of Γ such that (α_i, α_{i+1}) is a (directed) edge in Γ. A directed graph Γ is said to be *strongly connected* if for any two vertices α and β there exist directed paths from α to β as well as from β to α for all vertices α and β.

Define $\alpha \approx \beta$ if and only if there is a directed path from α to β as well as from β to α. It is easy to see that the relation $\approx$ is an equivalence relation. In the case when Γ is the orbital graph of a G-orbital Δ, we can show that $\approx$ is also a G-congruence. Take a directed path

$$\alpha = \alpha_0, \alpha_1, \ldots, \alpha_n = \beta$$

from α to β in Γ and let $g \in G$. Then the sequence

$$\alpha^g = \alpha_0^g, \alpha_1^g, \ldots, \alpha_n^g = \beta^g$$

is a directed path from α^g to β^g in Γ because $(\alpha_i, \alpha_{i+1}) \in \Delta$ implies $(\alpha_i^g, \alpha_{i+1}^g) = (\alpha_i, \alpha_{i+1})^g \in \Delta^g = \Delta$. The $\approx$-classes are called the *strong components* of the orbital graph Γ.

A weakening of the definitions made above leads to the notion of connectedness of graphs.

Definition 5.6 A *path* from α to β in a graph Γ is a sequence of vertices
$$\alpha = \alpha_0, \alpha_1, \ldots, \alpha_n = \beta$$
such that either (α_i, α_{i+1}) is an edge or (α_{i+1}, α_i) is an edge for $i = 0, 1, \ldots, n-1$. A directed graph Γ is said to be *connected* if there is a path from α to β for all vertices α and β.

The relation $\approx$ defined after Definition 5.5 will still remain a G-congruence even if we weaken the definition of $\alpha \approx \beta$ to mean that there is a path from α to β. The congruence classes are then called the *components* of the orbital graph. Connectedness of orbital graphs allows us to draw very useful conclusions about the primitivity of the action of a group on a transitive space.

Theorem 5.7 (D. G. Higman 1967) *Let Ω be a transitive G-space. Then Ω is primitive if and only if for every orbital Δ of G except the diagonal orbital, the orbital graph of Δ is connected.*

Proof: Assume Ω is primitive. Then we have seen that the relation $\approx$ defined by declaring that $\alpha \approx \beta$ if and only if there is a path from α to β is a G-congruence. So it is either trivial or universal. Since Δ is not the diagonal orbital, $\Delta = (\alpha, \beta)^G$ where $\alpha \neq \beta$. Since $\alpha \approx \beta$ the relation $\approx$ is not trivial. Hence it must be universal, which implies that any two vertices are joined by a path. This means that the graph is connected.

Conversely, suppose that every non-trivial orbital graph is connected. Let Γ be a non-trivial block of Ω with $\alpha, \beta \in \Gamma, \alpha \neq \beta$. Then the orbital graph of the orbital $(\alpha, \beta)^G$ is connected. If Γ is not the whole of Ω then there must be some edge in that graph with one vertex inside Γ and the other outside. This is the same as saying that there exists $g \in G$ such that Γ contains exactly one of α^g and β^g. But this violates the assumption that Γ is a block. Hence $\Gamma = \Omega$, and therefore Ω is primitive. $\square$

Note that we cannot replace 'connected' by 'strongly connected' in the above theorem as the following theorem illustrates.

Theorem 5.8 *The group $G := \mathrm{Aut}(\mathbb{Q}, <)$ is primitive but the strong components of the orbital graph of the orbital $(1, 2)^G$ are the singleton subsets of $\mathbb{Q}$.*

Proof: We have already seen in Example 4(e) that a 2-homogeneous group is primitive. Since G is 2-homogeneous (cf. Eg. 3(j)) it is primitive.

For an alternative proof of this fact using the last theorem note that G has only two non-trivial orbitals (cf. Eg. 5(c)). Since the connectedness of one of them implies the connectedness of the other, by Theorem 5.7, it is enough to show that the orbital graph of the orbital $\Delta := \{(q_1, q_2) \mid q_1 < q_2\}$ is connected. Given any pair of distinct elements q_1, q_2 from $\mathbb{Q}$ either $(q_1, q_2) \in \Delta$ or $(q_2, q_1) \in \Delta$. So the orbital graph of Δ is connected.

To prove the remainder of the theorem, note that the orbital Δ defined above is the same as the orbital $(1, 2)^G$. Let Δ_α denote the set of all vertices of the orbital graph defined above which are strongly connected to α. Clearly $\alpha \in \Delta_\alpha$. If $\beta \in \Delta_\alpha$ for some $\beta \neq \alpha$ then there exists a directed path

$$\alpha = \alpha_0, \alpha_1, \ldots, \alpha_n = \beta$$

from α to β and a directed path

$$\beta = \beta_0, \beta_1, \ldots, \beta_m = \alpha$$

from β to α. But this implies that

$$\alpha = \alpha_0 < \alpha_1 < \ldots < \alpha_n = \beta$$

and also that

$$\beta = \beta_0 < \beta_1 < \ldots < \beta_m = \alpha,$$

which gives a contradiction. So Δ_α is a singleton set containing the element α alone and therefore the orbital graph is not strongly connected. $\square$

The arguments we have just used show that if Ω is a primitive G-space then either the orbital graphs are strongly connected or the strong components are just the singleton sets. But in case Ω is finite, the distinction between strong connectedness and connectedness collapses.

Lemma 5.9 *Let Ω be a finite transitive G-space and let Δ be a nontrivial orbital. Then every component of the orbital graph is strongly connected.*

Proof: It suffices to show that if $\alpha = \alpha_0, \alpha_1, \ldots, \alpha_n = \beta$ is a path then we have a directed path from α to β. The problem is that some of the edges in our path might be in the 'wrong' direction. Suppose that (α_{i+1}, α_i) is an edge. To obtain a directed path from α to β we start by finding a directed path $\alpha_i = \gamma_0, \gamma_1, \ldots, \gamma_m = \alpha_{i+1}$ from α_i to α_{i+1}. Since Ω is transitive, there exists $g \in G$ such that $\alpha_{i+1}^g = \alpha_i$. Set $\gamma_{j-1} := \alpha_{i+1}^{g^j}$. Since Ω is finite, for some m we have $\alpha_{i+1} = \alpha_{i+1}^{g^{m+1}} = \gamma_m$. Now, since $(\gamma_{j-1}, \gamma_j) = (\alpha_{i+1}, \alpha_i)^{g^j} \in \Delta$ for all j we get a directed path from α_i to α_{i+1}. The path

$$\alpha = \alpha_0, \alpha_1, \ldots, \alpha_i, \gamma_1, \ldots, \gamma_{m-1}, \alpha_{i+1}, \ldots, \alpha_n = \beta$$

has one less edge going in the 'wrong' direction. Proceeding in this way, we can get rid of all the edges that go in the 'wrong' direction and finally get a directed path from α to β. $\square$

We can now state the version of the Theorem 5.7 in the case when Ω is finite. The proof is immediate from the last lemma and Theorem 5.7.

Theorem 5.10 *Let Ω be a finite transitive G-space. If Ω is primitive then every non-trivial orbital graph is strongly connected.* $\square$

Exercises:

5(i) Set $\Omega := \{1, \ldots, n\}$. What are the suborbits of A_n acting on Ω in the natural way? Describe the suborbits and orbitals of A_n acting on $\Omega \times \Omega$.

5(ii) Describe the suborbits and orbitals of $GL(2, \mathbb{R})$ acting on $\mathbb{R}^2$.

5(iii) Fill in the details of the proof of Theorem 5.2.

5(iv) (Oxford Final Honour School 1972) Suppose that G acts transitively on sets Ω_1 and Ω_2. For $\alpha_i \in \Omega_i$ let $G_i := G_{\alpha_i}$. Show that
(a) the number of orbits of G_1 on Ω_2,
(b) the number of orbits of G_2 on Ω_1, and
(c) the number of orbits of G on $\Omega_1 \times \Omega_2$
are the same.
What is this number when G is S_n, and Ω_1 and Ω_2 are the sets of r-element and s-element subsets of $\{1, 2, \ldots, n\}$ respectively?

5(v) Let Ω be a transitive G-space, and let $\Delta := (\alpha, \beta)^G$ be an orbital of G.

    (a) Prove that if $G_{\alpha\beta} = \{g \in G \mid \alpha^g = \alpha \text{ and } \beta^g = \beta\}$ then the size of the suborbit $\Delta(\alpha)$ is equal to $|G_\alpha : G_{\alpha\beta}|$.

    (b) The set $\Delta^* = (\beta, \alpha)^G$ is called the *paired orbital* of Δ, and $\Delta^*(\alpha)$ is the suborbit paired to $\Delta(\alpha)$. Prove that if Ω is finite then $|\Delta(\alpha)| = |\Delta^*(\alpha)|$.

    (c) Now think about the orbital graph with respect to Δ. The number of directed edges going out from a vertex is equal to $|\Delta(\alpha)|$ and the number of directed edges going into a vertex is equal to $|\Delta^*(\alpha)|$. Construct an example of a group G and a set Ω, such that there is an G-orbital Δ such that $|\Delta(\alpha)| \neq |\Delta^*(\alpha)|$. [Hint: The set Ω must be infinite. One way to approach this problem is to think about what the orbital graph must look like.]

5(vi) Let $\Omega := \{1, \ldots, n\}$ and define C_n be the (cyclic) group generated by the permutation $(1, 2, \ldots, n)$.

    (a) Suppose n is a prime number. Draw the orbital graph with respect to $(1, 2)^{C_n}$ and $(1, 3)^{C_n}$.

    (b) Suppose $n = 2k$. Draw the orbital graphs with respect to $(1, 2)^{C_n}$ and $(1, k + 1)^{C_n}$.

5(vii) Determine the suborbits and the orbitals of the action of S_n on the sets $\Omega^{\{2\}}$ and $\Omega^{(2)}$ as described in Example 3(h) and Exercise 4(iii). Use the D. G. Higman criterion (as stated in Theorem 5.7) to obtain alternative proofs of primitivity (or imprimitivity) of these actions.

5(viii) A binary relation $\preceq$ is called a *pre-order* if $\alpha \preceq \beta$ and $\beta \preceq \gamma$ implies $\alpha \preceq \gamma$, and $\alpha \preceq \alpha$ holds for all α. Every equivalence relation is also a pre-order and a non-strict partial ordering $\leq$ is also a pre-order. A trivial pre-order is one for which $\alpha \preceq \beta$ if and only if $\alpha = \beta$. A G-space Ω is said to be *strongly primitive* if there is no G-invariant non-trivial pre-order.

    (a) Show that the G-space Ω is strongly primitive if and only if for every proper subset Σ and every choice of distinct points α, β from Ω there is an element $g \in G$ such that $\alpha^g \in \Sigma$ and $\beta^g \notin \Sigma$.

    (b) Show that the G-space Ω is strongly primitive if and only if for every choice of distinct points α, β the orbital graph with respect to the orbital $(\alpha, \beta)^G$ is strongly connected.

(c) Prove that every finite primitive G-space is strongly primitive.

(d) Give an example of a primitive G-space that is not strongly primitive.

Chapter 6

More about Symmetric Groups

We have already defined the symmetric and alternating groups in Chapter 2 and have also seen some of their basic properties. We will study some more properties of these groups in this chapter.

Recall that the symmetric group on a set Ω is the set of all permutations on Ω and is denoted by $\text{Sym}\,(\Omega)$. For a set Ω of finite size n we denote $\text{Sym}\,(\Omega)$ and $\text{Alt}\,(\Omega)$ by S_n and A_n respectively, and call them the symmetric and alternating groups of degree n respectively. Since there are $n!$ permutations on a finite set of size n, the symmetric group of degree n has $n!$ elements. The following theorem tells us the size of $\text{Sym}\,(\Omega)$ when Ω is infinite.

Theorem 6.1 *For an infinite set Ω, set $\lambda := |\Omega|$. Then*

$$|\text{Sym}\,(\Omega)| = 2^\lambda.$$

Proof: Let $\wp(\Omega)$ denote the power set of Ω. Then $|\wp(\Omega)| = 2^\lambda$. Also

$$
\begin{aligned}
|\text{Sym}\,(\Omega)| &\leq |\{g : \Omega \to \Omega\}| \\
&\leq |\{g : \Omega \to \wp(\Omega)\}| \\
&= (2^\lambda)^\lambda \\
&= 2^\lambda
\end{aligned}
$$

as λ is an infinite cardinal. Therefore $|\text{Sym}\,(\Omega)| \leq 2^\lambda$.

To prove the reverse inequality, let us write Ω as a disjoint union of 2-element subsets of Ω. That is, let $\Omega = \dot{\bigcup}_{i \in I}\{\alpha_i, \beta_i\}$. For every subset J of I, define a permutation $t_J := \prod_{i \in J}(\alpha_i \beta_i)$. Then different subsets J, J' of I will produce different permutations $t_J, t_{J'}$. This is because if $j \in I$ is such that $j \in J \setminus J'$ then $\alpha_j^{t_J} = \beta_j$ but $\alpha_j^{t_{J'}} = \alpha_j$. Since $|I| = \lambda$, there are 2^λ such subsets of I. So there are at least that many permutations in $\mathrm{Sym}\,(\Omega)$. $\square$

6.1   Normal subgroups of symmetric groups

As we have mentioned earlier in Section 2.4, the finite symmetric groups have the alternating groups as normal subgroups, and the alternating group A_n is simple, for $n \neq 4$ (see Thm. 2.13). It is also a fact that A_n is the only non-trivial proper (sub)normal subgroup of S_n, except when $n = 4$. So, when $n \neq 4$,

$$\{1\} \trianglelefteq A_n \trianglelefteq \mathrm{Sym}\,(\Omega)$$

is the only composition series for S_n. Let us now study the normal subgroups of symmetric groups on infinite sets.

Definition 6.2 For $g \in \mathrm{Sym}\,(\Omega)$, define the *support* of g to be

$$\mathrm{supp}(g) := \{\omega \in \Omega \mid \omega^g \neq \omega\}.$$

Thus $\mathrm{supp}(g)$ is the set of points of Ω actually moved by g. We call the size of $\mathrm{supp}(g)$, namely $|\mathrm{supp}(g)|$, the *degree* of the permutation g. The following facts are near immediate consequences of the definition:

1. $\mathrm{supp}(g) = \mathrm{supp}(g^{-1})$;

2. $\mathrm{supp}(fg) \subseteq \mathrm{supp}(f) \cup \mathrm{supp}(g)$, because a point moved by fg must be moved by either f or g; and

3. $\mathrm{supp}(f^{-1}gf) = (\mathrm{supp}(g))^f$. This is because

$$\omega \in \mathrm{supp}(f^{-1}gf) \Leftrightarrow \omega^{f^{-1}gf} \neq \omega \Leftrightarrow \omega^{f^{-1}g} \neq \omega^{f^{-1}} \Leftrightarrow (\omega')^g \neq \omega',$$

where $(\omega')^f = \omega$. But this is equivalent to $\omega' \in \mathrm{supp}(g)$ which in turn means $\omega \in \mathrm{supp}(g)^f$.

EXERCISE:

6(i) Suppose f and g are permutations such that

$$|\text{supp}(f) \cap \text{supp}(g)| = 1.$$

Prove that then the commutator $f^{-1}g^{-1}fg$ is a 3-cycle.

Let $\aleph_0 \leq k \leq n = |\Omega|$. Note here that we are using k and n to denote infinite cardinals. Define

$$\text{BS}(\Omega, k) := \{g \in \text{Sym}\,(\Omega) \mid |\text{supp}(g)| < k\}.$$

Using the facts above, it is easy to prove that $\text{BS}(\Omega, k)$ is a normal subgroup of $\text{Sym}\,(\Omega)$. It is called the *bounded symmetric group* on Ω bounded by k.

The group $\text{BS}(\Omega, \aleph_0)$ is the group of all those permutations of Ω that have finite support. These permutations are called *finitary permutations*. Set

$$\text{FS}(\Omega) := \text{BS}(\Omega, \aleph_0).$$

The group $\text{FS}(\Omega)$ is called the *finitary symmetric group* on Ω.

A finitary permutation can be written as a product of a finite number of transpositions. Therefore, it makes sense to talk about even or odd finitary permutations. Define

$$\text{Alt}\,(\Omega) := \{g \in \text{FS}(\Omega) \mid g \text{ is even}\}.$$

This group is called the *(finitary) alternating group* on Ω. Using methods similar to those used for finite alternating groups it can be proved that $\text{Alt}\,(\Omega)$ is generated by 3-cycles, is a normal subgroup of $\text{Sym}\,(\Omega)$, and also that it is simple. It has index 2 in $\text{FS}(\Omega)$.

We end this section by stating a theorem which gives us a list of all the normal subgroups of $\text{Sym}\,(\Omega)$ when Ω is infinite.

Theorem 6.3 *Let N be a normal subgroup of* $\text{Sym}\,(\Omega)$. *Set* $n := |\Omega|$. *Then N is one of the groups in the chain*

$$\{1\} \trianglelefteq \text{Alt}\,(\Omega) \trianglelefteq \text{FS}(\Omega) \trianglelefteq \text{BS}(\Omega, \aleph_1) \trianglelefteq \ldots \trianglelefteq \text{BS}(\Omega, n) \trianglelefteq \text{Sym}\,(\Omega). \quad \square$$

We have already seen that all the groups in the above list are indeed normal subgroups of $S := \mathrm{Sym}\,(\Omega)$. We can show that there are no more by proving that the normal closure of any element g in S (which is defined to be the smallest normal subgroup of S containing g) is precisely one of the groups already in the list. See Schreier & Ulam (1933) for a proof of the theorem for countable Ω and Baer (1934) for a generalisation to arbitrary infinite sets. In fact, Baer (1934) proves a stronger result, namely that even if N is a subnormal subgroup of $\mathrm{Sym}\,(\Omega)$ it must be one of the groups in the above chain. Also refer to Bertram (1973) and Droste & Göbel (1979) for more on the subject.

6.2   Groups with finitary permutations

Throughout this section we shall assume G to be a permutation group on a set Ω. We have already observed in the last section that alternating groups contain all the 3-cycles and are generated by 3-cycles. We shall use that fact in the proof of the following theorem.

Theorem 6.4 (Wielandt 1960) *If G is primitive and contains a 3-cycle, then* $\mathrm{Alt}\,(\Omega) \leq G$.

Jordan proved the theorem for finite Ω in Note C of his book *Traité des substitutions* (1870).

Proof: Let G contain the 3-cycle $(\omega_1\ \omega_2\ \omega_3)$. A subset Γ of Ω will be said to be 'good' if

1. $\omega_1, \omega_2, \omega_3 \in \Gamma$ and

2. $\mathrm{Alt}\,(\Gamma) \leq G$. This means that for every element $h \in \mathrm{Alt}\,(\Gamma)$ there is an element $g \in G$ such that $\gamma^h = \gamma^g$ for all $\gamma \in \Gamma$, and $\alpha^g = \alpha$ for all $\alpha \in \Omega \setminus \Gamma$.

In particular, the set $\{\omega_1, \omega_2, \omega_3\}$ is good.

     Suppose that Γ_1 and Γ_2 are 'good' and let $(\alpha\ \beta\ \gamma)$ be a 3-cycle with $\alpha, \beta, \gamma \in \Gamma_1 \cup \Gamma_2$. If all three of α, β and γ lie in either Γ_1 or Γ_2 then $(\alpha\ \beta\ \gamma)$ lies in G. So let us suppose that $\alpha, \beta \in \Gamma_1$ and $\gamma \in \Gamma_2 \setminus \Gamma_1$. Then some conjugate of $(\alpha\ \beta\ \gamma)$ by an element g of $\mathrm{Alt}\,(\Gamma_1)$ is $(\omega_1\ \omega_2\ \gamma)$ which lies in $\mathrm{Alt}\,(\Gamma_2)$ (for example, if $\alpha, \beta, \omega_1, \omega_2$ are all different then the element g can be taken to be $(\alpha\ \omega_1)(\beta\ \omega_2)$).

Thus $g^{-1}(\alpha\beta\gamma)g \in \mathrm{Alt}\,(\Gamma_2)$. Therefore $(\alpha\beta\gamma) \in G$. But $\mathrm{Alt}\,(\Gamma_1 \cup \Gamma_2)$ is generated by 3-cycles. Hence $\Gamma_1 \cup \Gamma_2$ is 'good'.

Consequently, if $\Gamma' := \bigcup \{\Gamma \mid \Gamma$ 'good'$\}$ then it follows that Γ' is 'good'. To see this, take any even finitary permutation g of Γ'. Since its support is finite there are finitely many 'good' sets $\Gamma_1, \Gamma_2, \ldots, \Gamma_k$ whose union contains that support. But then by what has just been proved, that union is a 'good' set, so $g \in G$. Thus $\mathrm{Alt}\,(\Gamma') \leq G$, whence Γ' is 'good'.

Now, suppose $\Gamma' \neq \Omega$. Then there exists $g \in G$ such that $\Gamma'^g \neq \Gamma'$ but $\Gamma'^g \cap \Gamma' \neq \emptyset$. This is because Γ' cannot be a block. Replace g by g^{-1} if $\Gamma'^g \subset \Gamma'$. So, in either case, we may assume that $\Gamma'^g \setminus \Gamma' \neq \emptyset$.

But $\Gamma'^g \cup \Gamma'$ is 'good'. This is because

$$\mathrm{Alt}\,(\Gamma'^g) = g^{-1}\mathrm{Alt}\,(\Gamma')g \leq G$$

and since $\Gamma'^g \cap \Gamma' \neq \emptyset$ we must have

$$\mathrm{Alt}\,(\Gamma'^g \cup \Gamma') = \langle \mathrm{Alt}\,(\Gamma'^g), \mathrm{Alt}\,(\Gamma') \rangle \leq G,$$

where the first equality follows from an argument similar to that used to show that $\Gamma_1 \cup \Gamma_2$ is 'good'. But this contradicts the maximality of Γ'. $\square$

Corollary 6.5 *Under the hypothesis of the last theorem, if Ω is infinite then G is highly transitive.* $\square$

Theorem 6.6 (Neumann 1976) *Suppose that all G-orbits in Ω are infinite. If Γ, Δ are finite subsets of Ω then there exists $g \in G$ such that $\Gamma^g \cap \Delta = \emptyset$.*

Proof: The proof is by induction on $k := |\Gamma|$. If $k = 0$ then there is nothing to prove. If $k = 1$ then the theorem follows from the assumption that all orbits are infinite.

So let us assume as inductive hypothesis (IH) that the theorem is true for sets Γ' of size $(k-1)$ and for arbitrary finite sets Δ'. By taking suitable translations of Γ by elements of G if necessary, we can assume that $\Gamma \not\subseteq \Delta$. Choose $\gamma_0 \in \Gamma \setminus \Delta$ and let $\Gamma_0 := \Gamma \setminus \{\gamma_0\}$. We then apply the (IH) to Γ_0. We can find elements $g_0, g_1, \ldots, g_d$ where $d := |\Delta|$ such that

$$\Gamma_0^{g_0} \cap \Delta \;=\; \emptyset$$
$$\Gamma_0^{g_1} \cap (\Delta \cup \Delta^{g_0}) \;=\; \emptyset$$
$$\dots$$
$$\Gamma_0^{g_d} \cap (\Delta \cup \dots \cup \Delta^{g_{d-1}}) \;=\; \emptyset.$$

If $\gamma_0{}^{g_i} \notin \Delta$ for some i, then $\Gamma^{g_i} \cap \Delta = \emptyset$. So let us suppose that $\gamma_0{}^{g_0}, \gamma_0{}^{g_1}, \dots, \gamma_0{}^{g_d} \in \Delta$. Then since $|\Delta| = d$, there must exist p, q with $p > q$ such that $\gamma_0{}^{g_p} = \gamma_0{}^{g_q}$. Set $g := g_p g_q^{-1}$. Then $\gamma_0{}^g = \gamma_0 \notin \Delta$ and $\Gamma_0^g \cap \Delta = \emptyset$ because $\Gamma_0^{g_p} \cap \Delta^{g_q} = \emptyset$. Thus $\Gamma^g \cap \Delta = \emptyset$ and our proof is complete. $\square$

Replacing Δ by Γ in the above theorem, the following corollary is immediate.

Corollary 6.7 *Suppose that all G-orbits in Ω are infinite. If Γ is a finite subset of Ω then there exists $g \in G$ such that $\Gamma^g \cap \Gamma = \emptyset$.* $\square$

Theorem 6.8 (Wielandt's Theorem) *Suppose that Ω is infinite, G is primitive and there exists a non-identity element $g \in G$ with $\mathrm{supp}(g)$ finite. Then*

$$\mathrm{Alt}\,(\Omega) \leq G.$$

Remark: Jordan (1871) showed that there is a function $f : \mathbf{N} \to \mathbf{N}$ such that if $|\Omega| > f(|\mathrm{supp}(g)|)$ then $\mathrm{Alt}\,(\Omega) \leq G$. For a proof of the theorem as stated above see Wielandt (1960). The proof we give below is due to John D. Dixon.

Proof: Let $\alpha \in \Sigma := \mathrm{supp}(g)$. Then $g \notin G_\alpha$ and so we must have $\langle G_\alpha, g \rangle = G$. This is because since G is primitive, the subgroup G_α is maximal in G (cf. Thm. 4.7 (iv)). If there exists a G_α-orbit Γ disjoint from Σ, then it must be G-invariant, which is a contradiction. So every G_α-orbit meets Σ. Since Σ is finite, there are only finitely many G_α-orbits.

If there was a non-trivial finite G_α-orbit, then by Lemma 5.3 it would follow that Ω is finite, which is not true. Therefore, every non-trivial G_α-orbit is infinite. Then by Corollary 6.7 there must exist

$x \in G_\alpha$ such that $\Sigma \cap \Sigma^x = \{\alpha\}$. Set $h := x^{-1}gx$. Then it follows that

$$\text{supp}(g) \cap \text{supp}(h) = \{\alpha\}.$$

Applying Exercise 6(i), it is easy to see that $k := g^{-1}h^{-1}gh$ is the 3-cycle $(\alpha \; \alpha^h \; \alpha^g)$. Then an application of Theorem 6.4 completes the proof of the theorem. □

Combining Theorem 6.3 with the last theorem, we get the following corollary.

Corollary 6.9 *If Ω is infinite and G is a primitive group of finitary permutations then $G = \text{Alt}\,(\Omega)$ or $G = \text{FS}(\Omega)$.* □

We end the chapter with an easy application of the concepts introduced in this chapter.

Lemma 6.10 *Suppose that Ω is infinite, that $G \le \text{FS}(\Omega)$ and that G is transitive. Then if ρ is a proper congruence, the ρ-classes are finite.*

Proof: Let us choose $\alpha \ne \beta$ in Ω such that $\alpha \not\equiv \beta \,(\text{mod}\,\rho)$. Since G is transitive we can choose a permutation $g \in G$ which takes α to β. But then $(\rho(\alpha))^g = \rho(\beta)$ where $\rho(\theta)$ is the ρ-class containing θ. This implies that $\rho(\alpha) \subseteq \text{supp}(g)$ and hence is finite. □

<u>EXERCISES:</u>

6(ii) Prove that if Ω is an infinite set of cardinality n then $\text{Sym}\,(\Omega)$ has 2^{2^n} subgroups.

6(iii) Show that if Ω is countably infinite then $\text{Sym}\,(\Omega)$ has $2^{\aleph_0}$ conjugacy classes of elements.

6(iv) How many conjugacy classes of elements does $\text{Sym}\,(\Omega)$ have when $|\Omega| = \aleph_1$, the cardinality of the continuum? Generalise this result.

6(v) Permutations of order 2 are known as *involutions*. Prove that if $|\Omega| > 2$ then every element of $\text{Sym}\,(\Omega)$ may be expressed as a product of two involutions.

6(vi) Let Ω be a finite set and let Ω_1, Ω_2 be two subsets such that $\Omega = \Omega_1 \cup \Omega_2$ and $\Omega_1 \cap \Omega_2 \neq \emptyset$. Let $\mathrm{Sym}\,(\Omega_1)$, $\mathrm{Sym}\,(\Omega_2)$ be identified with subgroups of $\mathrm{Sym}\,(\Omega)$ in the natural way. Prove that $\mathrm{Sym}\,(\Omega) = \langle \mathrm{Sym}\,(\Omega_1), \mathrm{Sym}\,(\Omega_2) \rangle$.

6(vii) Investigate what happens in Exercise 6(vi) when Ω is infinite.

(viii) Let G be a group acting on a set Ω and suppose that all orbits have cardinality $\geq n$ where n is some infinite cardinal number. Let Γ, Δ be subsets of Ω such that Γ is finite and $|\Delta| < n$. Prove that there exists $g \in G$ such that $\Gamma^g \cap \Delta = \emptyset$.

6(ix) Let Ω be any uncountable set. Exhibit a transitive permutation group G on Ω and a pair of countably infinite subsets Γ, Δ of Ω such that $\Gamma^g \cap \Delta \neq \emptyset$ for all $g \in G$.

6(x) Let G be a transitive subgroup of the group $\mathrm{BS}(\Omega, k)$ described in Section 6.1. Show that if ρ is any G-congruence and $|\rho|$ denotes the cardinality of any ρ-class then $|\rho| < k$.

6(xi) (Neumann 1976) Let G be a group of finitary permutations of an infinite set Ω and N be a normal subgroup of G. If G/N is finite then N is transitive on every infinite orbit of G in Ω.

6(xii) (Neumann 1976) If G is a transitive group of finitary permutations of an infinite set Ω, and if N is a transitive normal subgroup of G, then $N \geq G'$, where G' is the derived subgroup of G.

Chapter 7

Linear Groups

In this chapter we deal with groups of matrices and related concepts. We have already defined $GL(2, \mathbb{R})$ to be the group of all 2×2 invertible real matrices, and seen (in Examples 2(d) and 3(e)) how it acts on $\mathbb{R}^2$ by matrix multiplication on the right. We now generalise these ideas to arbitrary fields F and to larger matrices.

7.1 General Linear Groups

Given a field F, a matrix is a rectangular array with entries from F. We say that a matrix M is an $n \times m$ matrix if it has n rows and m columns.

We will be concerned mainly with square matrices. A square matrix with n rows is said to be of *degree n*. We write $M_n(F)$ for the set of all square matrices of degree n with coefficients in F. Square matrices of the same degree can be multiplied together. Let I_n be the square matrix of degree n that has ones along the principal diagonal and zeroes elsewhere (here zero and one denote the zero and identity element of the field F respectively). We call I_n the unit matrix of degree n. A square matrix is said to be *invertible* if it has a two-sided inverse. That is, a matrix $A \in M_n(F)$ is invertible if there exists a matrix $B \in M_n(F)$ such that $A \cdot B = I_n = B \cdot A$.

We can also define the *determinant* of square matrices. The determinant takes values from the field F. Square matrices with non-zero determinant are said to be *non-singular*. It is a well-known (and very useful) fact that a matrix is invertible if and only if it is non-singular.

Definition 7.1 Let F be a field. The *n-dimensional general linear group* GL(n, F) over the field F is defined to be the group of all invertible matrices of degree n with entries in the field F.

Since a product of invertible matrices is invertible it is easy to see that GL(n, F) is a group with matrix multiplication as the binary operation, and the unit matrix I_n as the identity element.

Then, as in the case of GL($2, \mathbb{R}$), the group GL(n, F) acts on the vector space F^n by right multiplication. We shall talk about this action in detail later. Given a vector space V over a field F, a *linear transformation* on V is a mapping $\phi : V \to V$ such that the equation

$$(\lambda x + \mu y)\phi = \lambda(x\varphi) + \mu(y\phi)$$

holds for all $\lambda, \mu \in F$ and $x, y \in V$. A linear transformation is *invertible* if it has an inverse. If V has dimension n (usually written as $\dim(V) = n$) then $V \cong F^n$. When the dimension n is finite then (invertible) matrices of degree n with entries in F correspond to (invertible) linear transformations of V. This gives us an alternative definition for the general linear group, namely

$$\text{GL}(V) := \text{GL}(n, F) = \{g \in \text{Sym}\,(V) \mid g \text{ is linear}\}.$$

Using this definition, we do not need n to be a finite cardinal number. For example, if $n = \aleph_0$, the group GL(n, F) is defined to be the group of all invertible linear transformations of the vector space V with countably infinite dimension over F.

EXERCISE:

7(i) Show that the group GL(V) acts transitively on the set of ordered bases of the vector space V.

Determinants are defined only for square matrices of finite degree. In this case the general linear group has a special subgroup.

Definition 7.2 The *n-dimensional special linear group* over the field
F is defined to be the group

$$SL(n, F) := \{M \in GL(n, F) \mid \det(M) = 1\},$$

where $\det(M)$ denotes the determinant of the matrix M.

This is a subgroup because

- $\det(A^{-1}) = \det(A)^{-1}$ and

- $\det(A \cdot B) = \det(A) \cdot \det(B)$.

In fact it is a normal subgroup. Matrices with determinant 1 are also
said to be *unimodular*.

The non-zero elements in a field F form a group F^*. The operation in this group is the same as multiplication in the field. Another
way of looking at the group $SL(n, F)$ is by looking at the homomorphism det : $GL(n, F) \rightarrow F^* = F \setminus \{0\}$ which maps a matrix in
$GL(n, F)$ to its determinant. Clearly the map is onto. The subgroup
$SL(n, F)$ is then the kernel of this map. Therefore, we must have
$GL(n, F)/SL(n, F) \cong F^*$.

Let us now look briefly at the special case when F is a finite field
with q elements where $q = p^n$ for some prime p. We then often write
$GL(n, q)$ and $SL(n, q)$ instead of $GL(n, F)$ and $SL(n, F)$ respectively.

Theorem 7.3

$$\begin{aligned}
|GL(n, q)| &= (q^n - 1)(q^n - q)\ldots(q^n - q^{n-1}) \\
&= q^{n(n-1)/2}(q^n - 1)(q^{n-1} - 1)\ldots(q - 1)
\end{aligned}$$

and

$$|SL(n, q)| = |GL(n, q)|/(q - 1).$$

Proof: Counting the number of elements in $GL(n, q)$ is the same
as counting the number of invertible matrices of degree n that can
be written down with elements from the q-element field. The way
to guarantee invertibility is to ensure that the row vectors (or the
column vectors) of the matrix form a linearly independent set.

So when we write down the first row in our matrix, we have $q^n - 1$ choices (because the first row can be any vector except the zero vector $\hat{0} := (0, 0, \ldots, 0)$). Given the first row, there are exactly q vectors which are scalar multiples of it and hence linearly dependent on it. Excluding these, we have $q^n - q$ choices for the second row.

Suppose we have written down the first r rows as $l_1. l_2, \ldots. l_r$. For the $(r + 1)$-th row, we can choose any vector which is not a linear combination of these r vectors. There are exactly q^r vectors which are linearly dependent on the given r vectors, as these vectors are of the form $\alpha_1 l_1 + \alpha_2 l_2 + \ldots + \alpha_r l_r$, with α_i from the field F. Therefore we have $q^n - q^r$ choices for the $(r + 1)$-th row. This completes the proof of the first part of the theorem.

The second part follows from the fact that

$$|\mathrm{GL}(n, q) : \mathrm{SL}(n, q)| = |F^*| = q - 1. \quad \square$$

Exercise:

7(ii) Exhibit a subgroup of $\mathrm{SL}(n, q)$ whose order is $q^{n(n-1)/2}$. [Hint: Consider the set of upper-triangular matrices with ones along the diagonal.]

Let us now look once again at the action by right multiplication of $\mathrm{GL}(n, F)$ on the vector space $V = F^n$ over F. More explicitly, given an element $x \in V$ and an element $M \in \mathrm{GL}(n, F)$ define the image of x under the action of M to be $xM \in F^n$. This defines an action because matrix multiplication is associative and because the identity element in $\mathrm{GL}(n, F)$ is the unit matrix. An element $M \in \mathrm{GL}(n, F)$ can also be considered to be a vector space automorphism of V. This is because the equation

$$(\lambda x + \mu y)M = \lambda(xM) + \mu(yM)$$

holds for all $\lambda, \mu \in F$, $x, y \in V$ and for all $M \in \mathrm{GL}(V)$.

It is easy to see that $\mathrm{GL}(V)$ is not transitive in its action on V (cf. Eg. 3(e)). But it is transitive on the set $V \setminus \{\hat{0}\}$, as it is transitive on ordered bases of V (cf. Ex. 7(i)). By an argument similar to that used in Example 3(i), we can show that it is not 2-transitive on $V \setminus \{\hat{0}\}$, except when the field F has only 2 elements. Since V is assumed to have dimension at least 2, the transitive group $\mathrm{GL}(V)$ acting on $V \setminus \{\hat{0}\}$ is imprimitive if $|F| > 2$, because the non-zero elements of 1-dimensional subspaces form blocks of imprimitivity of this system.

7.2 Projective groups

A matrix of the form λI_n where $\lambda \in F^*$ is called a *scalar matrix*. Clearly the group of all scalar matrices is isomorphic to F^*. It is easily seen to be a normal subgroup of $\mathrm{GL}(n, F)$. In fact, a stronger result is true.

Lemma 7.4 *The centre of* $\mathrm{GL}(n, F)$ *is the set of all scalar matrices. That is,*

$$Z(\mathrm{GL}(n, F)) = \{\lambda I_n \mid \lambda \in F^*\}.$$

Furthermore, $Z(\mathrm{SL}(n, F)) = Z(\mathrm{GL}(n, F)) \cap \mathrm{SL}(n, F).$ $\square$

EXERCISE:

7(iii) Let $K_n := \{\lambda \in F \mid \lambda^n = 1\}$, the multiplicative group of n-th roots of unity in the field F. Show that the group $Z(\mathrm{SL}(n, F))$ of scalar matrices in the special linear group is isomorphic to K_n.

The *n-dimensional projective general linear group* $\mathrm{PGL}(n, F)$ is defined to be the factor group $\mathrm{GL}(n, F)/Z(\mathrm{GL}(n, F))$ and the *n-dimensional projective special linear group* $\mathrm{PSL}(n, F)$ to be the factor group $\mathrm{SL}(n, F)/Z(\mathrm{SL}(n, F))$. Using the Second Isomorphism Theorem (Theorem 1.6), it follows that

$$\mathrm{PSL}(n, F) = \mathrm{SL}(n, F).Z(\mathrm{GL}(n, F))/Z(\mathrm{GL}(n, F)).$$

Given a vector space V of finite dimension $n \geq 2$ over the field F, we can similarly define the groups $\mathrm{PGL}(V)$ and $\mathrm{PSL}(V)$. We often refer to the groups $\mathrm{SL}(n, F)$ and $\mathrm{PSL}(n, F)$ even when it is not assumed that n is finite. If n is infinite, these refer to $\mathrm{GL}(n, F)$ and $\mathrm{PGL}(n, F)$ respectively.

One important property of the groups $\mathrm{PSL}(V)$ is the fact (which we state without further comment) that they are simple whenever

V is a vector space of finite dimension at least 3 over an arbitrary field F. Thus the linear groups, like the alternating groups, are an important source of simple groups. For more details about normal subgroups of $GL(n, F)$ when n is infinite see Rosenberg (1958).

EXERCISE:

7(iv) Prove that
  (a) $PGL(n, \mathbb{C}) = PSL(n, \mathbb{C})$ for every n;
  (b) $PGL(n, \mathbb{R}) \cong PSL(n, \mathbb{R})$ if and only if n is odd;
  (c) if $n > 1$ then $PGL(n, \mathbb{Q}) \not\cong PSL(n, \mathbb{Q})$.

Considering the action of $GL(V)$ on the vector space V we note that $GL(V)$ maps vector subspaces to vector subspaces of the same dimension. This allows us to make the following definitions.

Let $0 \leq k \leq n-1$. Define Σ_k to be the set of all $(k+1)$-dimensional subspaces of V. The group $GL(V)$ acts on V and maps a $(k + 1)$-dimensional subspace to another $(k + 1)$-dimensional subspace. So it follows that $GL(V)$ also acts on Σ_k. The following lemma is easy.

Lemma 7.5 *For $0 \leq k < n - 1$, the kernel of the action of $GL(V)$ on Σ_k is $Z(GL(V))$.* □

Since $PGL(V) = GL(V)/Z(GL(V))$, it also acts on Σ_k. Furthermore, from the above lemma it follows that its action is faithful. In fact, we can say more.

Lemma 7.6 *Given $k \geq m \geq 0$, the group $GL(V)$ (and also $PGL(V)$) is transitive on the collection of pairs of k-dimensional subspaces of V with m-dimensional intersection.*

Proof: Given two such pairs of k-dimensional subspaces, W_1, W_2 and W_1', W_2', we want to show that there exists $M \in GL(V)$ which maps W_i to W_i' for $i = 1, 2$. For each pair, start with a basis for the intersection, extend it to a basis for one space, then to a basis for the sum with vectors chosen from the other space, and finally to a basis for V. Since $GL(V)$ is transitive on ordered bases of V (cf. Ex. 7(i)), it follows that $GL(V)$ has the required properties as does $PGL(V)$. □

Lemma 7.7 *The group* $\mathrm{PGL}(V)$ *acts primitively on the set* Σ_k *for* $0 \leq k < n - 1$. *It is 2-transitive only when* $k = 0$ *and* $k = n - 2$. $\square$

Observe that if $W_1, W_2 \in \Sigma_k$ then the orbit of $\mathrm{PGL}(V)$ containing $\{W_1, W_2\}$ is determined by $\dim(W_1 \cap W_2)$. The proof of the first part of Lemma 7.7 then follows from the first exercise below. The proof of Lemma 7.6 can be used to prove the second part of the lemma.

EXERCISE:

7(v) Show that the orbital graph of every non-diagonal orbital of $\mathrm{PGL}(V)$ is connected.

We define $\mathrm{PG}(V)$ to be the set Σ_0 and call it the $(n-1)$-*dimensional projective space* over F. The elements in Σ_k can be viewed as subsets of $\mathrm{PG}(V)$ and are called the *(projective) k-flats*. The 0-flats are precisely the points of $\mathrm{PG}(V)$ and $\mathrm{PG}(V)$ is the unique $(n-1)$-flat of this projective space.

We have already seen that the group $\mathrm{GL}(V)$ has an action on $\mathrm{PG}(V)$ as does $\mathrm{PGL}(V)$. We refer to any G-space which is equivalent to the $\mathrm{PGL}(n, F)$-space Σ_0 as being an $(n-1)$-dimensional *projective geometry* $\mathcal{P}_n(F)$ over F. With only trivial modification we can talk about infinite dimensional projective linear groups and infinite dimensional projective spaces and geometries.

7.3 Affine groups

Given a field F, an element $b \in F^n$ and an element $M \in \mathrm{GL}(n, F)$ we define a mapping $T_{M,b} : F^n \to F^n$ by

$$T_{M,b}(x) = xM + b.$$

These mappings are called the *affine transformations* on the vector space $V := F^n$ and constitute the *affine (general linear) group* defined as

$$\mathrm{AGL}(V) := \{T_{M,b} \mid M \in \mathrm{GL}(V), b \in V\}.$$

A little calculation shows that $\mathrm{AGL}(V)$ is indeed a group. It is a subgroup of $\mathrm{Sym}(V)$.

Under this definition it is clear that the group $\mathrm{AGL}(V)$ has a natural action on V, which takes an element $x \in V$ to the element $T_{M,b}(x) \in V$ under the action of the element $T_{M,b} \in \mathrm{AGL}(V)$. It is transitive as the element $T_{I_n, y-x} \in \mathrm{AGL}(n, F)$ maps $x \in V$ to $y \in V$.

Define a mapping $\mathrm{AGL}(V) \rightarrow \mathrm{GL}(V)$ which takes $T_{M,b}$ to M. This map is an epimorphism and the kernel of this map is the set

$$T(V) := \{ T_{I_n, b} \mid b \in V \}$$

which is the set of all translations of V. Thus $T(V) \trianglelefteq \mathrm{AGL}(V)$ and $\mathrm{AGL}(V)/T(V) \cong \mathrm{GL}(V)$.

The group $\mathrm{GL}(V)$ acting on V as defined earlier is a subgroup of $\mathrm{AGL}(V)$ for it is the stabiliser of the zero element $\hat{0}$ in $\mathrm{AGL}(V)$. Therefore, the group $\mathrm{AGL}(V)$ is 2-transitive in its action on V. It is not 3-transitive unless $|F| = 2$. This is because $\mathrm{GL}(V)$, the stabiliser of the zero element, is transitive (but not 2-transitive unless $|F| = 2$) in its action on $V \setminus \{\hat{0}\}$ (cf. Sec. 7.1). The action of $\mathrm{AGL}(V)$ is also primitive as it is 2-transitive.

The group $T(V)$ of translations of V is isomorphic to V, and is therefore abelian. It acts transitively on V by vector addition. Indeed its action is regular in the sense of Definition 3.8 (see also Ex. 4(x)).

Identifying $M \in \mathrm{GL}(V)$ with $T_{M,0} \in \mathrm{AGL}(V)$, we see that every affine transformation $T_{M,b}$ can be uniquely expressed as a composite of a linear transformation $T_{M,0}$ followed by a translation $T_{I_n, b}$. In fact, the group $\mathrm{AGL}(V)$ is a semidirect product of $\mathrm{GL}(V)$ and $T(V)$, and therefore also of $\mathrm{GL}(V)$ by V.

We refer to any G-space Ω equivalent to the $\mathrm{AGL}(n, F)$-space F^n as being an n-dimensional *affine geometry* $\mathcal{A}_n(F)$ over the field F. Note that although V as a $\mathrm{GL}(V)$-space was not transitive, (since the zero vector was special in that it could not be moved to any other vector), it is transitive as an $\mathrm{AGL}(V)$-space (which is sometimes also called an *affine space*). Thus all points in an affine space look alike. Furthermore, affine spaces are not just transitive, they are also primitive.

<u>EXERCISES:</u>

7(vi) Show that two affine transformations $T_{M,b}$ and $T_{M',b'}$ are equal if and only if $M = M'$ and $b = b'$.

7(vii) Prove that
(a) $\mathrm{AGL}(1,2) = S_2$
(b) $\mathrm{AGL}(1,3) = S_3$
(c) $\mathrm{AGL}(1,4) = A_4$
(d) $\mathrm{AGL}(2,2) = S_4$
where $\mathrm{AGL}(n,q)$ is the group $\mathrm{AGL}(V)$ with $V = F^n$ for a finite field F with q elements.

7(viii) Let

$$A := \{(a_{ij}) \in \mathrm{GL}(n+1, F) \mid a_{11} = 1, a_{i1} = 0 \text{ for} \cdot i > 1\}.$$

Show that A is a subgroup of $\mathrm{GL}(n+1, F)$ isomorphic to $\mathrm{AGL}(n, F)$.

7.4 Projective and affine spaces

We generalise the concept of a linear transformation in the following definition.

Definition 7.8 A *semilinear transformation* of a vector space V over a field F is a mapping $\phi : V \to V$, such that there exists an automorphism σ of the field F, for which

$$(\lambda x + \mu y)\phi = (\lambda^\sigma)(x\phi) + (\mu^\sigma)(y\phi)$$

for all $\lambda, \mu \in F$ and $x, y \in V$.

Note that we get a linear transformation on taking σ to be the identity automorphism on F. The group of all semilinear transformations on V is called the *semilinear group* and denoted by $\Gamma\mathrm{L}(V)$ and the corresponding quotient by the centre is the *projective semilinear group* $\mathrm{P}\Gamma\mathrm{L}(V)$. The group $\mathrm{A}\Gamma\mathrm{L}(V)$ is defined in the same way as $\mathrm{AGL}(V)$ using semilinear transformations instead of linear transformations.

We have already met the projective and affine spaces in the last two sections. To define them from the geometrical point of view, we need to understand what we mean by an incidence structure. Roughly speaking, an *incidence structure* is a collection of things, some of which we call *points*, some others *lines*, yet others *planes* etc., together with a binary relation of *incidence* between certain pairs from the collection. The usual Euclidean space provides us with a nice intuitive example of an incidence structure with the usual points. lines and planes in Euclidean space. The incidence relation in this case implies 'is part of' or 'lies on'. Thus a point may lie on a line or a plane, a line may lie on a plane, and so on.

The *projective space* $PG(V)$ corresponding to a vector space V is an incidence structure whose points are the 1-dimensional subspaces of V, lines are the 2-dimensional subspaces, planes are the 3-dimensional subspaces and hyperplanes are $(n-1)$-dimensional subspaces of V. Incidence is given by inclusion. It is known that if $\dim(V) > 2$ then the full automorphism group of $PG(V)$ is $P\Gamma L(V)$ (see Samuel 1988).

An *affine space* $AG(V)$ corresponding to the vector space V is an incidence structure which has the vectors of V as its points. the lines are sets of the form $L + b$, where L is a 1-dimensional subspace of V and $b \in V$ (i.e. cosets of 1-dimensional subspaces), the planes are of the form $P + b$, where P is a 2-dimensional subspace of V and so on. Incidence is given by inclusion. If $\dim(V) > 2$ then the full automorphism group of the affine space is $A\Gamma L(V)$, acting naturally on V.

EXERCISES:

7(ix) Show that the group $GL(V)$ is normal in $\Gamma L(V)$, and that $PGL(V)$ is normal in $P\Gamma L(V)$.

7(x) Show that the group $\Gamma L(V)$ is the semidirect product of $GL(V)$ by the automorphism group of F.

Chapter 8

Wreath Products

In this chapter we shall see an application of most of the concepts introduced so far. Wreath product constructions are very important in the study of permutation groups for a variety of reasons, some of which we shall see later in this chapter.

8.1 Definition

Let C be an abstract group and D a group acting on a set Δ. Define

$$K := C^\Delta = \{f \mid f : \Delta \to C\}.$$

Then K is easily seen to be a group under pointwise multiplication defined as follows. Given $f_1, f_2 \in K$ and $\delta \in \Delta$ we have

$$(f_1 f_2)(\delta) := f_1(\delta) f_2(\delta).$$

The identity element 1_K of K maps every element of Δ to the identity element of C. Let us define an action of D on K (which we shall later call *conjugation*), which takes $f \in K$ to $f^d \in K$ for $d \in D$, by specifying that

$$f^d(\delta) := f(\delta d^{-1}).$$

This defines an action because

$$f^{d_1 d_2}(\delta) = f(\delta(d_1 d_2)^{-1}) = f(\delta d_2^{-1} d_1^{-1}) = f^{d_1}(\delta d_2^{-1}) = (f^{d_1})^{d_2}(\delta)$$

and $f^{1_D}(\delta) = f(\delta)$ where $f \in K, d_i \in D, \delta \in \Delta$ and 1_D is the identity element of D. Using Lemma 8.2 (i) or otherwise, it is easy to see

67

that the map $f \mapsto f^d$ is an automorphism of K. The action of D on K thus gives us a rule θ associating an automorphism of K, say θ_d, with every element $d \in D$. In fact, θ is a homomorphism because

$$f\theta_{d_1 d_2} = f^{(d_1 d_2)} = f^{d_1}\theta_{d_2} = f\theta_{d_1}\theta_{d_2}$$

for all $f \in K$ and all $d_1, d_2 \in D$. We thus have all the ingredients required to define the (external) semidirect product of K by D.

Definition 8.1 The *wreath product* W of C by D is defined to be the semidirect product of K by D. That is,

$$W := K \rtimes D = C^\Delta \rtimes D$$

is defined on the set $\{(f, d) \mid f \in K, d \in D\}$ with multiplication defined by

$$(f_1, d_1)(f_2, d_2) = (f_1 f_2^{d_1^{-1}}, d_1 d_2).$$

We usually write $W := C \operatorname{Wr}_\Delta D$ or simply $W := C \operatorname{Wr} D$ to denote that W is a wreath product of C by D. The group D is called the *top group*, the group C is called the *bottom group* and the group K is called the *base group* of the wreath product W. The factors ($\cong C$) of K are called the *co-ordinate subgroups* of K indexed by Δ. We can also directly prove that W is a group, using the following lemma.

Lemma 8.2 Let f, f_i, g denote elements of K and let d, d_i be elements of D. Also let 1_X denote the identity element of the group X. Then

(i) $(f_1 f_2)^d = f_1^d f_2^d$

(ii) $((f_1, d_1)(f_2, d_2))(f_3, d_3) = (f_1, d_1)((f_2, d_2)(f_3, d_3))$

(iii) $(f, d)(1_K, 1_D) = (f, d) = (1_K, 1_D)(f, d)$

(iv) $(f, d)^{-1} = ((f^{-1})^d, d^{-1}) = ((f^d)^{-1}, d^{-1})$

(v) $(1_K, d)^{-1}(f, 1_D)(1_K, d) = (f^d, 1_D)$

(vi) $(f, d)(g, 1_D)(f, d)^{-1} = (fg^{d^{-1}}f^{-1}, 1_D).$

Proof: The proofs are absolutely routine. We do the first, second and fifth parts as samples.

(i)

$$(f_1 f_2)^d(\delta) \;=\; (f_1 f_2)(\delta d^{-1})$$
$$= f_1(\delta d^{-1}) f_2(\delta d^{-1}) \;=\; f_1^d(\delta) f_2^d(\delta)$$
$$= (f_1^d f_2^d)(\delta).$$

(ii)

$$((f_1, d_1)(f_2, d_2))(f_3, d_3) \;=\; (f_1 f_2^{d_1^{-1}}, d_1 d_2)(f_3, d_3)$$
$$= (f_1 f_2^{d_1^{-1}} f_3^{(d_1 d_2)^{-1}}, d_1 d_2 d_3) \;=\; (f_1 f_2^{d_1^{-1}} f_3^{d_2^{-1} d_1^{-1}}, d_1 d_2 d_3).$$

Similarly,

$$(f_1, d_1)((f_2, d_2)(f_3, d_3)) \;=\; (f_1, d_1)(f_2 f_3^{d_2^{-1}}, d_2 d_3)$$
$$= (f_1(f_2 f_3^{d_2^{-1}})^{d_1^{-1}}, d_1 d_2 d_3) \;=\; (f_1 f_2^{d_1^{-1}} f_3^{d_2^{-1} d_1^{-1}}, d_1 d_2 d_3);$$

using (i). Hence they are equal.

(v)

$$(1_K, d)^{-1}(f, 1_D)(1_K, d) = (1_K, d^{-1})(f, d) = (f^d, 1_D). \quad \square$$

Let us identify K and D with the subsets $\{(f, 1_D) \mid f \in K\}$ and $\{(1_K, d) \mid d \in D\}$ of W. Then it is easy to prove that they are subgroups of W. Furthermore, the last part of the last lemma shows that $K \trianglelefteq W$. As $D \cap K = \{(1_K, 1_D)\}$ and $K.D = W$ under the above identification, we see that W is indeed also the (internal) semidirect product of the groups K by D as defined at the end of Chapter 1.

This identification allows us to write elements of W as fd instead of (f, d). Then by part (v) of Lemma 8.2, we have $d^{-1} f d = f^d$. We can therefore think of the action of D by K as *conjugation*. The rule for multiplication can then be re-written as

$$(fd)(gd') = f(dgd^{-1})dd' = fg^{d^{-1}} dd'.$$

8(i) Define the *diagonal T* of the base group K of the wreath product
$W = C \operatorname{Wr}_\Delta D$ to be the set of all constant functions. That is,

$$T := \{f \in K \mid f(\delta) = f(\delta') \text{ for all } \delta, \delta' \in \Delta\}.$$

Show that $T \leq K$ and also that there is a natural isomorphism
that can be defined from T to C.

8.2   Wreath products as permutation groups

In the special case when the groups C and D act on sets, their wreath
product W also has an action on the cartesian product set. To see
this consider the wreath product $W = C \operatorname{Wr}_\Delta D$ where both C and
D are permutation groups acting on sets Γ and Δ respectively. The
group W then has a natural action on the set $\Gamma \times \Delta$ given by

$$(\gamma, \delta)(f, d) = (\gamma f(\delta), \delta d).$$

Note that since $f \in K$ we have $f(\delta) \in C$ so that the above equation
makes sense. It is easy to check that this actually defines an action.
Furthermore, a lot can be said about the nature of the action of W
on the set $\Gamma \times \Delta$ from the actions of C and D on Γ and Δ respectively
as the following theorem will show.

Theorem 8.3 (i) *If C acts faithfully on Γ and D acts faithfully on
Δ then W acts faithfully on $\Gamma \times \Delta$.*

(ii) *If C acts transitively on Γ and D acts transitively on Δ then W
acts transitively on $\Gamma \times \Delta$.*

Proof: (i) Let $(f, d) \in W$ be distinct from the identity. Then either
$f \neq 1_K$ or $d \neq 1_D$. In the first case, choose δ so that $f(\delta) \neq 1_C$ and
let $\gamma \in \Gamma$ be such that $\gamma f(\delta) \neq \gamma$. In the second case, let $\delta \in \Delta$ be
such that $\delta d \neq \delta$ and let $\gamma \in \Gamma$ be arbitrary. In both cases, we have
$(\gamma, \delta)(f, d) \neq (\gamma, \delta)$.

(ii) Given two pairs (γ_1, δ_1) and (γ_2, δ_2) choose $c \in C$ and $d \in D$
such that $\gamma_1 c = \gamma_2$ and $\delta_1 d = \delta_2$. Choose $f \in K$ such that $f(\delta_1) = c$.

Then the element $(f, d) \in W$ will take (γ_1, δ_1) to (γ_2, δ_2). □

We also note that the wreath product construction is 'associative' in the restricted sense that if G_1, G_2, G_3 are groups acting on sets $\Omega_1, \Omega_2, \Omega_3$ respectively, then the groups $(G_1 \operatorname{Wr}_{\Omega_2} G_2) \operatorname{Wr}_{\Omega_3} G_3$ and $G_1 \operatorname{Wr}_{\Omega_2 \times \Omega_3} (G_2 \operatorname{Wr}_{\Omega_3} G_3)$ are equal as permutation groups acting on the set $\Omega_1 \times \Omega_2 \times \Omega_3$. This makes iteration possible in wreath product constructions.

8.3 Imprimitivity of wreath products

Suppose that C and D act transitively on the sets Γ and Δ respectively, so that the wreath product $W = C \operatorname{Wr}_\Delta D$ acts transitively on $\Gamma \times \Delta$ under the action defined earlier as $(\gamma, \delta)(f, d) = (\gamma f(\delta), \delta d)$. This action is *imprimitive* in a natural way as we shall see in the following lemma.

Lemma 8.4 *Suppose* $|\Gamma|, |\Delta| > 1$. *Then the relation* ρ *defined on* $\Gamma \times \Delta$ *by*

$$(\gamma_1, \delta_1) \equiv (\gamma_2, \delta_2) \bmod \rho \Leftrightarrow \delta_1 = \delta_2$$

is a W-*congruence.*

Proof: Clearly ρ is an equivalence relation on $\Gamma \times \Delta$. That it is a W-congruence follows from the fact that $\delta_1 = \delta_2 \Leftrightarrow \delta_1 d = \delta_2 d$ for all $d \in D$. □

Notes:

I A ρ-class is of the form

$$\Gamma_\delta := \{(\gamma, \delta) \mid \gamma \in \Gamma\}$$

and hence is identifiable with Γ. Furthermore, W acts on each ρ-class in the same way as the group C acts on Γ.

II There are as many ρ-classes as there are elements in Δ. The base group K is the kernel of the action of W on the set of ρ-classes, because every element of K fixes every ρ-class setwise. Also since $W/K \cong D$, it follows that the action of W on the set of ρ-classes is identifiable with the action of D on Δ.

III Let G be transitive on Ω and let ρ be a G-congruence with Γ as a ρ-class. Then as we have seen before in Section 3.3 the setwise and pointwise stabilisers of Γ in G. namely $G_{\{\Gamma\}}$ and $G_{(\Gamma)}$ respectively are subgroups of G with $G_{(\Gamma)} \trianglelefteq G_{\{\Gamma\}}$. The factor group $G_{\{\Gamma\}}/G_{(\Gamma)}$ is the group of permutations induced by $G_{\{\Gamma\}}$ on Γ. Furthermore, since G is transitive on Ω and Γ is a congruence class, this factor group will be transitive on Γ. This is because, if $\gamma_1, \gamma_2 \in \Gamma$ and $g \in G$ maps γ_1 to γ_2 then g belongs to $G_{\{\Gamma\}}$. This factor group is called the *factor group produced by the congruence ρ* and is independent of the ρ-class Γ we choose. In our case of the wreath product W and the congruence ρ defined in the last lemma, C is the factor group produced by ρ on a congruence class Γ_δ.

One of the most important properties of wreath products is their universality as embedding groups for imprimitive groups. We show this in the following theorem (cf. Neumann 1976).

Theorem 8.5 *Let G be any transitive but imprimitive group of permutations on a set Ω and let ρ be a G-congruence on Ω. Let Γ be a ρ-class and let C be the factor group produced on Γ by the congruence ρ (cf. Note III above). Let Δ be the set Ω/ρ of all ρ-classes and let D be the group of permutations induced by G on Δ. Then Ω may be identified with $\Gamma \times \Delta$ in such a way that*

$$G \leq W := C \operatorname{Wr}_\Delta D.$$

Proof: To obtain an explicit embedding of G into W we choose a family $\{t_\delta\}_{\delta \in \Delta}$ of elements in G such that $\Gamma t_\delta = \delta$ for all $\delta \in \Delta$. Note that if $\delta_1 \neq \delta_2$ then $t_{\delta_1} \neq t_{\delta_2}$. The family $\{t_\delta\}_{\delta \in \Delta}$ is then a transversal for the group $G_{\{\Gamma\}}$ in G. Then the embedding of G into W can be described by

$$g \mapsto (f_g, g^\Delta)$$

where g^Δ is the permutation induced by g on Δ ($= \Omega/\rho$) and f_g is an element of the base group $K := C^\Delta$ such that

$$f_g(\delta) = (t_\delta g t_{\delta g}^{-1}).$$

This definition makes sense because $\Gamma(t_\delta g t_{\delta g}^{-1}) = \delta g t_{\delta g}^{-1} = \Gamma$, which implies that $(t_\delta g t_{\delta g}^{-1})$ lies in $G_{\{\Gamma\}}$. We now show that this map has the desired properties.

We first show that the map is a homomorphism. To show that

$$(f_{gh}, (gh)^\Delta) = (f_g, g^\Delta)(f_h, h^\Delta)$$

we have to show that $f_{gh} = f_g f_h^{g^{-1}}$ and that $(gh)^\Delta = g^\Delta.h^\Delta$. The second equality holds because the map $g \mapsto g^\Delta$ is a homomorphism. The first equality follows from

$$f_g f_h^{g^{-1}}(\delta) = f_g(\delta)f_h(\delta g) = (t_\delta g t_{\delta g}^{-1})(t_{\delta g} h t_{\delta gh}^{-1}) = (t_\delta g h t_{\delta gh}^{-1}) = f_{gh}(\delta).$$

To show that it is a monomorphism it is enough to show that the kernel of the map is trivial. If g is in the kernel then this implies that the element (f_g, g^Δ) is the identity element in W. But this means that g^Δ induces the identity element on Δ and then $f_g(\delta) = t_\delta g t_\delta^{-1}$ is the identity element of C, for every $\delta \in \Delta$. This in turn implies that g acts as the identity on Γ as well. Thus g fixes every point of Ω and is hence the identity element of G.

Finally to get an embedding, we need to identify Ω with $\Gamma \times \Delta$. The inverse of the mapping $(\gamma, \delta) \mapsto \gamma t_\delta$ will do. To see this, we only have to show that $\gamma_1 t_{\delta_1} = \gamma_2 t_{\delta_2}$ implies $\gamma_1 = \gamma_2$ and $t_{\delta_1} = t_{\delta_2}$. Now $\gamma_1, \gamma_2 \in \Gamma$, implies $\gamma_1 t_{\delta_1}$ and $\gamma_2 t_{\delta_2}$ belong to the same ρ-class, so that $t_{\delta_1} = t_{\delta_2}$ and therefore $\delta_1 = \delta_2$, and so also $\gamma_1 = \gamma_2$. $\square$

EXERCISE:

8(ii) If C and C_1 are permutation groups on the same set Γ, and D and D_1 are permutation groups on the same set Δ such that $C \leq C_1$ and $D \leq D_1$ then $C \operatorname{Wr}_\Delta D \leq C_1 \operatorname{Wr}_\Delta D_1$.

The following corollary is immediate from Theorem 8.5 and the last exercise.

Corollary 8.6 *The permutation group G defined in Theorem 8.5 is such that*

$$G \leq \operatorname{Sym}(\Gamma) \operatorname{Wr}_\Delta \operatorname{Sym}(\Delta). \quad \square$$

Thus we see that just as the symmetric groups are the universal embedding groups for arbitrary permutation groups, all imprimitive permutation groups can be naturally embedded in wreath products of symmetric groups of suitable size. For other applications of wreath products see Wells (1976), Möller (1991) and Bhattacharjee (1994, 1995).

8.4   Variations of wreath products

Let us consider the wreath product $W = C \operatorname{Wr}_\Delta D$ as defined earlier where C and D are permutation groups acting on sets Γ and Δ respectively. We have already seen how this group has an imprimitive action on the set $\Gamma \times \Delta$. We can define other actions too, as we shall see by defining an action of W on the set Γ^Δ. The wreath product W has a *product action* on this set defined in the following way. For $\omega \in \Gamma^\Delta$ and $(f, d) \in W$ define $\omega^{(f,d)}$ to be that element of Γ^Δ which maps $\delta \in \Delta$ to the element $\omega(\delta d^{-1}) f(\delta d^{-1}) \in \Gamma$. It is not too difficult to prove that this actually defines an action on the set Γ^Δ. What is surprising, however, is that this action can be primitive under certain conditions as we have the following theorem (cf. Dixon & Mortimer 1996, Lem. 2.7A).

Theorem 8.7 *Let C and D be permutation groups acting on Γ and Δ respectively. If $|\Gamma|, |\Delta| > 1$, then $W := C \operatorname{Wr}_\Delta D$ acts primitively on Γ^Δ under the product action defined above if and only if Δ is finite, D acts transitively on Δ and C acts primitively but not regularly on Γ.* $\square$

The following exercise is an illustration of the last theorem.

<u>EXERCISE:</u>

8(iii) Consider the natural action of $S_n \operatorname{Wr} C_2$ on the set of ordered pairs $\Omega^{(2)}$ from the set $\Omega := \{1, 2, \ldots, n\}$. Here S_n acts on the ordered pairs as defined in Exercise 4(iii) while the transposition in C_2 just interchanges the coordinates. Show that the action is primitive with two non-trivial orbitals.

The wreath product W we have defined in Section 8.1 is sometimes further qualified as the *permutational wreath product*. It is also sometimes called the *unrestricted wreath product* of the groups C by D to differentiate it from the *restricted wreath product* of the two groups which is defined using the restricted direct product (instead of the cartesian product) in the definition of K. Thus, in this case,

$$K := C^{(\Delta)} = \{f : \Delta \to C \mid \mathrm{supp}(f) \text{ is finite}\}$$

where $\mathrm{supp}(f) := \{\delta \in \Delta \mid f(\delta) \neq 1_C\}$. The restricted wreath product of C by D is usually denoted as $C \, \mathrm{wr}_\Delta \, D$ or simply as $C \, \mathrm{wr} \, D$. Clearly the unrestricted and restricted versions coincide when Δ is a finite set.

Wreath product constructions are often used to build interesting abstract groups. In particular, this occurs with $\Delta = D$. This is possible since every group acts on itself via the right regular representation. In this case the action of D on K is simply right multiplication. The wreath product thereby defined is called the *standard wreath product* of C by D. Much more can be said about the structure of standard wreath products by utilising the group structure of the set Δ. This subject has been studied extensively in Neumann (1964). We state some of the results in that paper as exercises at the end of this chapter.

Many other variations to the theme of wreath products have also been constructed. A generalisation of the wreath products called the *twisted wreath product* of groups is studied in B. H. Neumann (1963), and is also described in Suzuki (1982). Another construction, called *wreath towers* is described in Hall (1962). *Generalised wreath products* have been studied by Holland (1969).

EXERCISES:

8(iv) Show that a restricted wreath product is a subgroup of the corresponding unrestricted wreath product.

8(v) (Neumann 1964) Let W denote the unrestricted standard wreath product of C by D as defined earlier. Let T be the diagonal subgroup of the base group as defined in Exercise 8(i).
   (a) Given a subgroup H of G, the *centraliser* of H in G is defined to be the set of all those elements of G which commute

with every element of H. Show that the centraliser of D in
W is $Z(D) \times T$, where $Z(D)$ denotes the centre of D.

(b) Show that if C is not trivial then the centre of W is the
centre of T.

(c) If neither C nor D is trivial and if

$$W = C \operatorname{Wr} D = P \times Q$$

then show that one of P, Q is central in W (that is, is
contained in the centraliser of W), and the other contains D.

(d) Given subgroups A, B of a group G, we say that A is a
complement of B in G if $A \cap B = \{1\}$ and $A.B = G$. Show
that any two complements for K in G are conjugate.

(e) If $W = C \operatorname{Wr} D$ and $W' = C' \operatorname{Wr} D'$ and $W \cong W'$, then
show that $C \cong C'$ and $D \cong D'$.

Chapter 9

Rational Numbers

We have already seen some properties of the rational numbers and its group of automorphisms in Examples 3(j), 5(c) and Theorem 5.8. In this chapter we look at the set of rational numbers in greater detail to discover many fascinating properties of its group of order-automorphisms.

9.1 Cantor's Theorem

Definition 9.1 A *linearly ordered set* or a *totally ordered set* is a non-empty set A equipped with a binary relation $<$ satisfying:

(i) for all $x \in A$, we have $x \not< x$ (irreflexivity);

(ii) for all $x, y \in A$, we have $x < y \Rightarrow y \not< x$ (anti-symmetry);

(iii) for all $x, y, z \in A$, $x < y$ and $y < z \Rightarrow x < z$ (transitivity);

(iv) for $x, y \in A$, either $x < y, x = y$ or $x > y$ (linearity).

A non-empty set A satisfying conditions (i), (ii) and (iii) is called a *partially ordered set*. Note that (i) and (iii) imply (ii). Moreover, a partially ordered set A is said to be *dense* if

(v) for $x, y \in A$ with $x < y$, there exists $z \in A$ such that $x < z < y$.

It is said to be *unbounded* or *without endpoints* if

(vi) for $x \in A$, there exist $y, z \in A$ such that $y < x < z$.

We note that $(\mathbf{Q}, <)$, where $\mathbf{Q}$ is the set of rational numbers and $<$ is the usual order on $\mathbf{Q}$, is a countable dense linear order without endpoints.

Definition 9.2 Let $(A, <)$ and $(B, <)$ be two partially ordered sets. A map $\theta : A \to B$ is an *order-isomorphism* if θ is bijective and order-preserving, i.e. for all $x, y \in A$, $x < y \Leftrightarrow x\theta < y\theta$.

EXERCISES:

9(i) Show that $\mathbf{N}, \mathbf{Z}, \mathbf{Q}$ are not order-isomorphic.

9(ii) Given totally ordered sets A and B, their *sum* $A + B$ denotes a totally ordered set whose domain is the disjoint union of isomorphic copies of A and B, with the natural ordering induced on each copy, and with all elements of the copy of A less than all elements of the copy of B. Let $\mathbf{N}^*$ denote the set $\mathbf{N}$ with the reverse order. Show that $\mathbf{N}^* + \mathbf{N}$ is order-isomorphic to $\mathbf{Z}$ but not to $\mathbf{N} + \mathbf{N}^*$.

9(iii) Show that $\mathbf{Q}$ is order-isomorphic to its reverse $\mathbf{Q}^*$.

9(iv) Partially ordered sets which are order-isomorphic are said to belong to the same *order type*. Show that all order types of the same finite size are order-isomorphic.

9(v) An order type is said to be *rigid* if its group of automorphisms is trivial. Show that order types of finite size are rigid. Also show that $\mathbf{N}$ is rigid and $\mathbf{Z}$ just admits itself as automorphism group, acting regularly.

Theorem 9.3 (Cantor's Theorem) *Any two countable dense linearly ordered sets without endpoints are order-isomorphic. Equivalently, if $(A, <)$ satisfies* (i) – (vi) *of Definition 9.1 and A is countable, then there is an order-isomorphism $\theta : \mathbf{Q} \to A$.*

Proof: Let $(A, <)$ be a countable ordered set satisfying (i) – (vi) of Definition 9.1. Let $\mathbf{Q} = \{q_0, q_1, \ldots\}$ and $A = \{a_0, a_1, \ldots\}$ be enumerations of $\mathbf{Q}$ and A respectively. We define a map $\theta : \mathbf{Q} \to A$ step by step.

Step 0: Put $q_0\theta := a_0$.

Step $(k+1)$: Suppose that we have defined θ for $q_0, \ldots, q_k$ so that θ is order-preserving on them and now want to define $q_{k+1}\theta$. We must specify the ordering on $q_0, \ldots, q_k$ induced from $\mathbb{Q}$. So let $q_{i_0} < q_{i_1} < \ldots < q_{i_k}$ where $i_l \in \{0, 1, \ldots, k\}$.

    Case 1: $q_{i_0} < \ldots < q_{i_k} < q_{k+1}$. Then let s be the smallest integer such that $q_{i_0}\theta < \ldots < q_{i_k}\theta < a_s$ (which exists due to (vi)), and put $q_{k+1}\theta := a_s$.

    Case 2: $q_{k+1} < q_{i_0} < q_{i_1} < \ldots < q_{i_k}$. Then proceed in a manner similar to Case 1.

    Case 3: $q_0 < \ldots < q_{i_l} < q_{k+1} < q_{i_{l+1}} < \ldots < q_{i_k}$. We have $q_{i_l}\theta < q_{i_{l+1}}\theta$ in A and therefore, by (v), we can find the least integer s such that $q_{i_l}\theta < a_s < q_{i_{l+1}}\theta$. We put $q_{k+1}\theta := a_s$.

Proceeding in this manner, in the end we will have defined θ on the whole of A. This mapping is clearly injective (simply because all inequalities are strict). It is order-preserving as it is so at each step. We now show that θ is surjective. Suppose that θ is not so. Let m be the smallest integer such that a_m is not in $\mathrm{Im}\,(\theta)$. We can choose r such that

$$\{a_0, a_1, \ldots, a_{m-1}\} \subseteq \{q_0\theta, q_1\theta, \ldots, q_r\theta\}.$$

Reordering, we have $q_{i_0} < q_{i_1} < \ldots < q_{i_r}$, where $i_l \in \{0, 1, \ldots, r\}$. Let $b_l = q_{i_l}\theta \in A$. Then $b_0 < b_1 < \ldots < b_r$. Suppose

$$b_0 < b_1 < \ldots < b_l < a_m < b_{l+1} < \ldots < b_r.$$

If t is the smallest integer such that

$$q_{i_l} < q_t < q_{i_{l+1}}$$

(which exists by (v)), then when we were defining $q_t\theta$ we must have set it equal to a_m. This contradicts the fact that $a_m \notin \mathrm{Im}\,(\theta)$. Similarly, the cases when $a_m < b_0$ and $b_r < a_m$ can be dealt with using (vi). We therefore conclude that θ is onto. This completes the proof of the theorem. $\square$

The argument used in the above proof of Cantor's Theorem for constructing the order-isomorphism, is called *'going forth'*. The construction ensured that the map is one-to-one and order-preserving. Subsequently the surjectivity of the map had to be proved. Since there is complete symmetry in what we know about the sets $\mathbb{Q}$ and A we can build some symmetry into the argument for a proof as well. This is what we do when we use a *'back and forth'* argument as demonstrated below. The main advantage is that surjectivity also is guaranteed by the construction and no extra work need be done to prove it.

Alternative proof of Cantor's Theorem: We build the isomorphism θ up by steps. As before, choose enumerations of the sets $\mathbb{Q}$ and A. At an odd step we fix the image of an element of $\mathbb{Q}$ and at an even step we fix the preimage of an element of A.

Step 0: Define $q_0\theta := a_0$.

Step $(2i + 1)$ (odd step): Let θ be already defined on a finite set $M \subseteq \mathbb{Q}$. Let j be the least integer such that θ is not yet defined on q_j. We find the least integer k such that defining $q_j\theta := a_k$, we get θ as an order-preserving map from $M \cup \{q_j\}$.

Step $(2i + 2)$ (even step): Let $N \subseteq A$ be the range of θ up to Step $(2i + 1)$. Let j be the least integer such that $a_j \notin N$. We find the least integer k such that defining $q_k\theta := a_j$, we get θ as an order-preserving map on $N\theta^{-1} \cup \{q_k\}$.

The mapping θ defined on $\mathbb{Q}$ is clearly bijective and order-preserving and this completes the proof of the theorem. $\square$

<u>EXERCISE:</u>

9(vi) (a) Describe explicitly an order-isomorphism $\mathbb{Q} \longrightarrow \mathbb{Q} \setminus \{0\}$;
      (b) Show that $\mathbb{R}$ is not order-isomorphic to the set $\mathbb{R} \setminus \{0\}$.
      [This exercise illustrates the fact that Cantor's Theorem does not generalise to uncountable sets.]

9.2   Back and forth *vs* going forth

The *'back and forth'* construction is a very basic and flexible tool for building isomorphisms and automorphisms. It is essential to the proof of Fraïssé's Theorem which we shall discuss in Chapter 14. This technique is also much stronger and can be applied to many other situations where a *'going forth'* argument does not suffice. We illustrate this fact by an example.

For that we first need to prove the existence of dense codense colourings of the rationals. By this we mean colouring the rationals in two colours, say red and blue, in such a way that between any two points there is a blue point as well as a red point. To do that we use some known properties of the rationals. We know that the set of dyadic fractions, that is, rationals of the form $a/2^m$ for some $m \in \mathbf{N}$ and some $a \in \mathbf{Z}$, is dense in the rationals. Let us colour all the dyadic fractions red and all the other rationals blue. Then it is not a very difficult exercise to show that this gives us a dense codense colouring of the rationals. We can now proceed with the example (cf. Cameron 1990, Sec. 5.2).

EXAMPLE: Let $(A, <)$ be a countable dense linearly ordered set without endpoints. Then, by Cantor's Theorem, $(A, <)$ is order-isomorphic to $(\mathbf{Q}, <)$, and therefore has a dense subset whose complement is also dense. We call the points in the given subset red, and those in its complement blue. So, between any two points there is a blue point as well as a red point. Using a 'back and forth' argument it can be proved that any two such sets are order-isomorphic where the red points are mapped onto the red points and the blue points are mapped onto the blue points.

However in this case the 'going forth' argument does not suffice. It is easier to illustrate this by talking about automorphisms of a structure, rather than isomorphisms between structures. We show that for two enumerations of our structure A, one may not get an automorphism of A by the 'going forth' argument. In fact, given any second enumeration, it is possible to choose a first enumeration in such a way that the map defined by 'going forth' is not onto. This is what we will show below.

Let $\{b_0, b_1, \ldots\}$ be any (second) enumeration of A. We find a (first) enumeration $\{a_0, a_1, \ldots\}$ of A such that for the mapping θ defined by 'going forth,' we have $b_0 \notin \text{Im}(\theta)$.

Step 0: Let r be the least integer such that b_r has the opposite colour to that of b_0. Find an a_0 in A having the same colour as that of b_r such that b_0 lies between a_0 and b_r. Put $a_1 := b_0$. Clearly then $a_0\theta = b_r$ and $a_1\theta \neq b_0$.

Step n: At this stage let us assume that $a_0, a_1, \ldots, a_{n-1}$ have already been specified in such a way that when θ is defined on them with the 'going forth' argument, b_0 is not the image of any of the points. Now, let k be the smallest integer such that $b_k \notin \{a_0, a_1, \ldots, a_{n-1}\}$. We may suppose that

$$a_{l_0} < a_{l_1} < \ldots < a_{l_{n-1}},$$

where $l_i \in \{0, 1, \ldots, n-1\}$. Let $b_0 \in (a_{l_j}\theta, a_{l_{j+1}}\theta)$.

Case 1: If $b_k \notin (a_{l_j}, a_{l_{j+1}})$ or if b_k has opposite colour to that of b_0, then set $a_n := b_k$ to guarantee that $a_n\theta \neq b_0$.

Case 2: Suppose that $b_k \in (a_{l_j}, a_{l_{j+1}})$ and that b_k and b_0 have the same colour. Let b_m be the first point in $(a_{l_j}\theta, b_0)$ having the opposite colour to b_0, and c any point in the interval $(b_k, a_{l_{j+1}})$ having same colour as b_m. Set $a_n := c$ and $a_{n+1} := b_k$. Then $a_n\theta = b_m$ and $a_{n+1}\theta = b_k\theta \neq b_0$.

Similarly the cases when $b_0 < a_{l_0}$ and $a_{l_{n-1}} < b_0$ can be dealt with.

At the end we shall have obtained an enumeration $\{a_0, a_1, \ldots\}$ of A such that b_0 never gets into the image of θ.

Exercise:

9(vii) For any fixed $k \in \{1, 2, \ldots, \aleph_0\}$, show that there is up to isomorphism a unique countable totally ordered set with k colours, each dense and without endpoints.

9.3   Order-automorphisms of the rationals

The set $G := \text{Aut}(\mathbb{Q}, <)$ of all the order-automorphisms of $\mathbb{Q}$ is a group which acts on $\mathbb{Q}$. We have already seen in Example 3(j) that

G is transitive (but not 2-transitive) and highly homogeneous on $\mathbb{Q}$. Furthermore, Theorem 5.8 tells us that G is primitive.

Clearly every open interval of $\mathbb{Q}$, whether it is of type (p, q), $(-\infty, p)$ or (p, ∞) for any $p, q \in \mathbb{Q}$, has all the properties (i) – (vi) of Definition 9.1, and is hence order-isomorphic to $\mathbb{Q}$, by Cantor's Theorem. Also if any $g \in G$ fixes a point $p \in \mathbb{Q}$, then the restrictions of g to the intervals $(-\infty, p)$ and (p, ∞) give order-automorphisms defined on those intervals (cf. Eg. 3(j)). In the notation set up after Definition 3.12, we therefore have the following corollary.

Corollary 9.4 *Let $p \in \mathbb{Q}$. Define*

$$\mathbb{Q}_{<p} := \{q \in \mathbb{Q} \mid q < p\} \ and \ \mathbb{Q}_{>p} := \{q \in \mathbb{Q} \mid q > p\}.$$

Then the stabiliser G_p of p in G is such that

$$G_p = G^{\mathbb{Q}_{<p}} \times G^{\mathbb{Q}_{>p}} \cong \mathrm{Aut}\,(\mathbb{Q}, <) \times \mathrm{Aut}\,(\mathbb{Q}, <). \quad \square$$

Theorem 9.5 *The group G is highly homogeneous on $\mathbb{Q}$* $\quad \square$.

This has already been proved in Example 3(j). The argument given there illustrates Cantor's Theorem insofar as the sets A_i, B_i are countable dense linearly ordered sets without endpoints and the linear maps f_i are explicit examples of order-isomorphisms from A_i to B_i.

Definition 9.6 A group G acting on a set Ω is said to be *oligomorphic* in its action on Ω if G has finitely many orbits on Ω^k for every natural number k.

Note that the action of G on Ω^k is defined by

$$(\omega_1, \ldots, \omega_k)^g = (\omega_1^g, \ldots, \omega_k^g).$$

There is a rich theory of oligomorphic groups, partly stimulated by its connection with model theory based on the fact that any oligomorphic group is dense in the automorphism group of an $\aleph_0$-categorical structure (cf. Defn. 14.1), and conversely. We recommend Cameron (1990) for a fascinating study of oligomorphic permutation groups.

Lemma 9.7 *Let $\Omega^{\{k\}}$ denote the set of all k-element subsets of Ω. Then G is oligomorphic in its action on Ω if and only if it has finitely many orbits on $\Omega^{\{k\}}$ for every natural number k.*

Proof: Let $\phi : \Omega^k \to \wp(\Omega)$ be the map

$$(\omega_1, \omega_2, \ldots, \omega_k) \mapsto \{\omega_1, \omega_2, \ldots, \omega_k\}.$$

Clearly, ϕ is a G-morphism, and its image is $\bigcup_{j=1}^{k} \Omega^{\{j\}}$. Since orbits map to orbits under a G-morphism (cf. Ex. 3(i)), if there are only finitely many G-orbits in Ω^k then there are only finitely many orbits in $\bigcup_{j=1}^{k} \Omega^{\{j\}}$, and therefore also only finitely many in $\Omega^{\{k\}}$. Conversely, since each element of $\bigcup_{j=1}^{k} \Omega^{\{j\}}$ is the image under ϕ of at most $k!$ elements of Ω^k. the pre-image under ϕ of each G-orbit in $\bigcup_{j=1}^{k} \Omega^{\{j\}}$ is the union of at most $k!$ orbits in Ω^k. Therefore. if there are only finitely many orbits in $\bigcup_{j=1}^{k} \Omega^{\{j\}}$ then there are only finitely many in Ω^k. □

The following corollary is immediate from the last lemma and Theorem 9.5.

Corollary 9.8 *The group $\mathrm{Aut}(\mathbb{Q}, <)$ is oligomorphic on $\mathbb{Q}$.* □

EXERCISES:

9(viii) Let Ω be a countable set and consider the natural action of $\mathrm{Sym}(\Omega)$ on the set $\Omega^{\{2\}}$ of unordered pairs from Ω (see Example 3(h)). Show that the action is oligomorphic.

9(ix) Show that if G is oligomorphic then any subgroup of finite index, and any point stabiliser is also oligomorphic.

9(x) Show that the infinite dihedral group (cf. Eg. 5(e)) is not oligomorphic on $\mathbb{Z}$.

9(xi) Show that if G is a countably infinite group and $\mathrm{Aut}(G)$ is oligomorphic on G, then
   (a) there is $d \in \mathbb{N}$ such that all elements of G have order at most d; and
   (b) there is a mapping $f : \mathbb{N} \to \mathbb{N}$ such that for any finite $A \subset G$, we have $|\langle A \rangle| \leq f(|A|)$.

9(xii) Show that $\mathrm{GL}(V)$ is oligomorphic on V, if V is an $\aleph_0$-dimensional vector space over a finite field.

We end this chapter with a theorem which tells us about normal subgroups of $\mathrm{Aut}\,(\mathbb{Q}, <)$ (cf. Glass 1981, Thm. 2.3.2). As before, set $G := \mathrm{Aut}\,(\mathbb{Q}, <)$ and define

$$
\begin{aligned}
B &:= \{g \in G \mid \mathrm{supp}(g) \text{ is bounded}\}, \\
R &:= \{g \in G \mid \mathrm{supp}(g) \text{ is bounded below}\}, \\
L &:= \{g \in G \mid \mathrm{supp}(g) \text{ is bounded above}\}.
\end{aligned}
$$

[A mnemonic is that elements of R live on the *right* and those of L live on the *left*.] With this notation, we have the following theorem.

Theorem 9.9 *The sets B, R and L are normal subgroups of G, and they are the only non-trivial proper subnormal subgroups of G. Furthermore, $B = L \cap R$ and $G = L \cdot R$.* □

<u>Exercises:</u>

9(xiii) Given a set $\Sigma \subseteq \mathbb{Q}$ and $g \in G := \mathrm{Aut}\,(\mathbb{Q}, <)$, define the set Σ^g to be a *transform* of Σ under G. Show that Σ has $< 2^{\aleph_0}$ transforms under G if and only if Σ is a union of finitely many intervals with rational end points (or $\pm\infty$).
[Hint: Define $H := G_{\{\Sigma\}}$. Show that
(a) all H-orbits are convex;
(b) there are only finitely many H-orbits; and
(c) the end points of the H-orbits are rational.]

9(xiv) Show that if a is the translation $\xi \mapsto \xi + 1$ and $f \in \mathrm{Aut}\,(\mathbb{Q}, <)$ then there exist conjugates f_1 and f_2 of a such that $f = f_1 f_2^{-1}$.

9(xv) A **Z**-sequence is a sequence $\{\xi_n\}_{n \in \mathbb{Z}}$ of rationals such that $\xi_n < \xi_{n+1}$ for all n and such that $\xi_n \to \pm\infty$ as $n \to \pm\infty$. Show that the group $\mathrm{Aut}\,(\mathbb{Q}, <)$ acts transitively on the set of **Z**-sequences.

9(xvi) (Truss 1989) If $G := \mathrm{Aut}\,(\mathbb{Q}, <)$ and $H \leq G$ is such that $|G : H| < 2^{\aleph_0}$ then show that there is a finite set $\Delta \subseteq \mathbb{Q}$ such that $H = G_{(\Delta)}$. [Hint: Use Ex. 9(xiii) to get started.]

9(xvii) Show that the derived group B' of B is simple. where B is the group defined just before Theorem 9.9. What can you say about the simplicity of B? [Hint: See Lem. 6.4 of Glass (1981).]

Chapter 10

Jordan Groups

In this chapter, we introduce the notion of a Jordan set, and study some properties of groups containing Jordan sets. Most of the material covered in this chapter can be found in Adeleke & Neumann (1996a).

10.1 Definitions and some examples

Definition 10.1 Let Ω be a G-space and let $\Omega = \Delta \cup \Gamma$ be a partition of Ω with $|\Gamma| > 1$. If there exists a subgroup H of G that fixes every point of Δ and is transitive on Γ then Γ is called a *Jordan set* for G in Ω and Δ is called a *Jordan complement*.

For $k \in \mathbf{N}$, if G is $(k+1)$-fold transitive and $|\Delta| \leq k$ then the set $\Gamma = \Omega \setminus \Delta$ is automatically a Jordan set. We call such Jordan sets *improper*. Otherwise, they are called *proper* Jordan sets.

Definition 10.2 If G is transitive on Ω and there is a proper Jordan set for G in Ω then we call G a *Jordan group*.

Notes:

I If Γ is a Jordan set, and if $H \leq G_{(\Omega \setminus \Gamma)}$ is a subgroup of G that fixes every point of $\Omega \setminus \Gamma$ and is transitive on Γ then H is called a group *associated* with the Jordan set Γ. Alternatively, we can say that Γ is a Jordan set *with respect to H*.

II We say that Γ is a *primitive* Jordan set if H is (or may be chosen to be) primitive on Γ. We use the same rule to define *k-transitive, k-homogeneous* or *k-primitive* Jordan sets. We can define *highly transitive* Jordan sets similarly.

To illustrate the definitions just made and to convince ourselves that many of the familiar groups we know are Jordan groups, we mention some examples here. Many more examples will be given in Chapters 11, 12 and 15.

Examples :

10(a) Any set of size ≥ 2 in Ω is a Jordan set in $\text{Sym}(\Omega)$ or in $\text{BS}(\Omega, k)$. For $\text{Alt}(\Omega)$ any set of size ≥ 3 will do. These are examples of highly transitive Jordan sets when Ω is infinite.

10(b) The projective and affine groups acting on points in projective and affine space respectively, are natural sources of Jordan groups (cf. Sec. 11.2).

10(c) Given a wreath product $H \operatorname{Wr} K$ acting transitively but imprimitively on the set $\Gamma \times \Delta$, it is easy to see that for $\delta \in \Delta$ the set $\{(\gamma, \delta) \mid \gamma \in \Gamma\}$ (which is a block of imprimitivity under the action) is a Jordan set. In fact wreath products are rather typical examples of imprimitive Jordan groups.

10(d) Any non-empty open interval in $\mathbb{Q}$ can be easily seen to be a Jordan set in $\text{Aut}(\mathbb{Q}, <)$ (compare Eg. 3(j) and Sec. 11.3).

10.2   Basic properties of Jordan sets

Lemma 10.3 *If* Γ *is a Jordan set and* $g \in G$ *then so is* Γ^g.

Proof: Let $H \leq G$ fix $\Omega \setminus \Gamma$ pointwise and be transitive on Γ. Then $g^{-1}Hg \leq G$ fixes $\Omega \setminus \Gamma^g$ pointwise and is transitive on Γ^g. $\square$

We need the following definition for the next lemma.

Definition 10.4 A family $\mathcal{F}$ of subsets of a set Ω is said to be *connected* if for $\Gamma_1, \Gamma_2 \in \mathcal{F}$ there exist $\Sigma_0, \Sigma_1, \ldots, \Sigma_l \in \mathcal{F}$ such that $\Sigma_0 = \Gamma_1, \Sigma_l = \Gamma_2$ and $\Sigma_i \cap \Sigma_{i+1} \neq \emptyset$ for $i = 0, 1, \ldots, l-1$.

Lemma 10.5 (i) *If* Γ_1, Γ_2 *are Jordan sets and* $\Gamma_1 \cap \Gamma_2 \neq \emptyset$, *then* $\Gamma_1 \cup \Gamma_2$ *is also a Jordan set.*

(ii) *The union of a connected family of Jordan sets is a Jordan set.*

Proof: (i) Clearly, there is nothing to prove if either $\Gamma_1 \subseteq \Gamma_2$ or $\Gamma_2 \subseteq \Gamma_1$. So we shall assume this is not the case. Let H_1 and H_2 be subgroups of G associated with the Jordan sets Γ_1 and Γ_2 respectively and $H := \langle H_1, H_2 \rangle$, the group generated by H_1 and H_2. Then we claim that H is transitive on $\Gamma_1 \cup \Gamma_2$ and fixes every point in its complement. The last assertion is clear since both H_1 and H_2 do so. To prove transitivity, the only case there could be a problem is when we have two points $\gamma_1 \in \Gamma_1 \setminus \Gamma_2$ and $\gamma_2 \in \Gamma_2 \setminus \Gamma_1$. But since $\Gamma_1 \cap \Gamma_2 \neq \emptyset$ we can move γ_1 to an element γ' of the intersection by an element $h_1 \in H_1$ and then move γ' to γ_2 by an element $h_2 \in H_2$. The element $h_1 h_2 \in H$ will then take γ_1 to γ_2.

(ii) The proof of this part follows from the first and is left to the reader as an exercise. $\square$

Corollary 10.6 *Suppose* G *is a Jordan group acting on a set* Ω.

(i) *Let* α *and* β *be distinct points of* Ω. *If there exists a Jordan set* Γ_0 *containing* β *but not* α, *then there is a unique Jordan set* Γ *containing* β *but not* α *and such that* Γ *is maximal subject to these conditions.*

(ii) *Suppose that* $\Gamma_0, \Delta_0 \subseteq \Omega$, $\Gamma_0 \cap \Delta_0 = \emptyset$ *and* Γ_0 *is a Jordan set. Then there is a Jordan set* Γ *such that* $\Gamma_0 \subseteq \Gamma$, $\Gamma \cap \Delta_0 = \emptyset$ *and* Γ *is maximal subject to this.*

Proof: (i) Take the family of all Jordan sets containing β but not α. Then it is a connected non-empty family. By the second part of the last lemma, the union of all the sets in that family is a Jordan set. Moreover it is also maximal subject to containing β but not containing α.

(ii) The second part is a generalisation of the first. The proof follows in exactly the same way. $\square$

Lemma 10.7 *Suppose that* G *is primitive on* Ω *and that* Γ *is a Jordan set. Then for all* $\alpha, \beta \in \Omega$ *there exists* $g \in G$ *such that* $\alpha, \beta \in \Gamma^g$.

Proof: Define a binary relation ρ on Ω by declaring that ω_1, ω_2 are ρ-equivalent if there exists $g \in G$ such that $\omega_1, \omega_2 \in \Gamma^g$. We prove that ρ is a G-congruence. It is certainly reflexive and symmetric. To show that it is transitive, suppose $\omega_1, \omega_2 \in \Gamma^g$ and $\omega_2, \omega_3 \in \Gamma^h$ for some $g, h \in G$. If $\omega_3 \in \Gamma^g$ then ω_1, ω_3 are ρ-equivalent. If not, since Γ^g is a Jordan set, there exists $k \in G$ fixing ω_3 and mapping ω_2 to ω_1. Then Γ^{hk} contains both ω_1 and ω_3. So ρ is an equivalence relation. Clearly ρ is G-invariant and hence ρ is a G-congruence. It is non-trivial, since $|\Gamma| \geq 2$. Since G is primitive, it must be universal. $\square$

Remark: There are plenty of examples to show that the above lemma does not work when G is imprimitive.

10.3   More properties of Jordan sets

Lemma 10.8 *Suppose that Ω is infinite and that G is primitive on Ω. Then*

(i) *if there is a finite Jordan set Γ then $\text{Alt}(\Omega) \leq G$;*

(ii) *if there are Jordan sets Γ_1, Γ_2 such that $\Gamma_1 \cap \Gamma_2$ is finite but non-empty then $\text{Alt}(\Omega) \leq G$.*

So, in particular, in both cases G is highly transitive. Also every subset Σ of Ω with more than two members is a Jordan set for G.

Proof: (i) This follows immediately from Theorem 6.8 as we can find a finitary permutation (which moves only elements in Γ) in G.

(ii) By Theorem 6.8, it suffices to exhibit a finitary permutation in G. If one of Γ_1, Γ_2 is finite then clearly a finitary permutation certainly exists. So let us suppose that both of them are infinite. Since their intersection is finite, both $\Gamma_1 \setminus \Gamma_2$ and $\Gamma_2 \setminus \Gamma_1$ are nonempty. Choose $\alpha \in \Gamma_1 \cap \Gamma_2, \beta \in \Gamma_1 \setminus \Gamma_2$ and $\gamma \in \Gamma_2 \setminus \Gamma_1$ and choose $x \in H_1, y \in H_2$ (where H_1 and H_2 are subgroups of G associated with Γ_1 and Γ_2 respectively) such that $\alpha^x = \beta$ and $\gamma^y = \alpha$. Define $g := x^{-1}y^{-1}xy$. Then

$$\beta^g = \beta^{x^{-1}y^{-1}xy} = \alpha^{y^{-1}xy} = \gamma^{xy} = \gamma^y = \alpha$$

so that $g \neq 1$. To show that g is finitary we need to find a finite subset Σ of Ω such that g fixes every point of $\Omega \setminus \Sigma$. We claim that the finite set

$$\Sigma := (\Gamma_1 \cap \Gamma_2) \cup (\Gamma_1 \cap \Gamma_2)^x \cup (\Gamma_1 \cap \Gamma_2)^y$$

will do. To see this note that g fixes every point in the complement of $\Gamma_1 \cup \Gamma_2$. So we need to consider only the following two cases for $\omega \in \Omega$.

Case (i): $\omega \in \Gamma_1 \setminus \Sigma$. Since $\omega \notin (\Gamma_1 \cap \Gamma_2)^x$ we have $\omega^{x^{-1}} \notin (\Gamma_1 \cap \Gamma_2)$. So $\omega^{x^{-1}}$ not in Γ_2. Thus $\left(\omega^{x^{-1}}\right)^{y^{-1}} = \omega^{x^{-1}}$ since y and hence y^{-1} acts only on Γ_2. Also $\omega^y = \omega$ as $\omega \notin \Gamma_2$. Therefore

$$\omega^g = \omega^{x^{-1}y^{-1}xy} = \left(\omega^{x^{-1}}\right)^{y^{-1}xy} = \left(\omega^{x^{-1}}\right)^{xy} = \omega^y = \omega.$$

Case(ii): $\omega \in \Gamma_2 \setminus \Sigma$. The proof of this case is similar to the first case using the fact that $\omega \notin (\Gamma_1 \cap \Gamma_2)^y$. □

Lemma 10.9 *Let* Γ_1, Γ_2 *be Jordan sets for* G *in* Ω. *Suppose that* $\Gamma_1 \cap \Gamma_2 \neq \emptyset$ *and let* $\Gamma := \Gamma_1 \cup \Gamma_2$. *Then*

  (i) *if* Γ_1, Γ_2 *are* k-*transitive, so is* Γ;

 (ii) *if* Γ_1, Γ_2 *are* k-*primitive, so is* Γ;

(iii) *if* Γ_1, Γ_2 *are* k-*homogeneous and* $|\Gamma_1|, |\Gamma_2| \geq 2k$, *then so is* Γ.

Proof: Let H_1 and H_2 be subgroups of G associated with Γ_1 and Γ_2 respectively and such that H_1, H_2 are k-transitive, k-primitive, or k-homogeneous as appropriate. The results are trivial if either $\Gamma_1 \subseteq \Gamma_2$ or $\Gamma_2 \subseteq \Gamma_1$. So let us assume that both $\Gamma_1 \setminus \Gamma_2$ and $\Gamma_2 \setminus \Gamma_1$ are nonempty and define $H := \langle H_1, H_2 \rangle$.

(i) We use induction on k. For $k = 1$, we use Lemma 10.5 to obtain the result. So assume $k \geq 2$, and that the result is true for $k - 1$. Choose $\alpha \in \Gamma_1 \setminus \Gamma_2$. Then $\Gamma_1 \setminus \{\alpha\}$ is a $(k-1)$-transitive Jordan set. So $(\Gamma_1 \setminus \{\alpha\}) \cup \Gamma_2$ is also $(k-1)$-transitive, by inductive hypothesis. But $(\Gamma_1 \setminus \{\alpha\}) \cup \Gamma_2 = \Gamma \setminus \{\alpha\}$ and H_α is associated with this Jordan set. Thus H_α is $(k-1)$-transitive on $\Gamma \setminus \{\alpha\}$ which means H is k-transitive on Γ, by Theorem 3.13. This completes the induction.

(ii) Let us first prove that Γ is primitive whenever Γ_1 and Γ_2 are primitive. Since H_1, H_2 are primitive on Γ_1, Γ_2 respectively, they are transitive. Hence H is transitive on Γ, by Lemma 10.5. Let ρ be a non-trivial H-congruence on Γ. Then $\rho_{|\Gamma_1}$ and $\rho_{|\Gamma_2}$ are H_1- and H_2-congruences respectively. If they are both trivial congruences then $\rho_{|\Gamma_1 \cap \Gamma_2}$ is also trivial, so that ρ-classes in $\Gamma_1 \cap \Gamma_2$ have size 1. But then all ρ-classes will have size 1, which is not possible. So one of $\rho_{|\Gamma_1}$ and $\rho_{|\Gamma_2}$ is non-trivial, say the first. But Γ_1 is primitive which implies that $\rho_{|\Gamma_1}$ is the universal relation. So the whole of Γ_1 is one ρ-class. If $\rho_{|\Gamma_2}$ were trivial, then ρ-classes in $\Gamma_2 \setminus \Gamma_1$ would have size 1, while Γ_1 is a ρ-class of size more than 1, in contradiction to the fact that H is transitive on Γ. Therefore $\rho_{|\Gamma_2}$ must also be the universal relation. Since $\Gamma_1 \cap \Gamma_2 \neq \emptyset$ it follows that the whole of Γ forms one ρ-class, that is, ρ is the universal relation on Γ. Therefore H is primitive on Γ.

The assertion for k-primitivity now follows by induction on k, exactly as in the case of k-transitivity.

(iii) Let Δ_0 be a k-element subset of $\Gamma_1 \cap \Gamma_2$, and let Δ be any k-subset of Γ. We assume first that $|\Gamma_1 \cap \Gamma_2| \geq k$. Then since H_1 is m-homogeneous for every $m \leq k$ (Theorem 3.19) we can choose h_1 from H_1 such that $(\Delta \cap \Gamma_1)^{h_1} \subseteq \Gamma_1 \cap \Gamma_2$. Then $\Delta^{h_1} \subseteq \Gamma_2$ and so there exists $h_2 \in H_2$ such that $\Delta^{h_1 h_2} = \Delta_0$. Thus H is transitive on k-sets in Γ.

Next, assume that $|\Gamma_1 \cap \Gamma_2| < k$. Choose $h_1 \in H$ such that $|\Delta^{h_1} \cap \Gamma_1|$ is as large as possible. We may certainly suppose that $\Delta^{h_1} \cap \Gamma_1 \subseteq \Gamma_1 \setminus \Gamma_2$. If $\Delta^{h_1} \not\subseteq \Gamma_1$ then choose γ from $\Delta^{h_1} \setminus \Gamma_1$ and $h_2 \in H_2$ such that $\gamma^{h_2} \in \Gamma_1 \cap \Gamma_2$. Clearly then,

$$|\Delta^{h_1 h_2} \cap \Gamma_1| > |\Delta^{h_1} \cap \Gamma_1|$$

and this contradicts our choice of h_1. Thus $\Delta^{h_1} \subseteq \Gamma_1$, and since H_1 is k-homogeneous, it follows that the k-set Δ can be moved to the preassigned k-subset Δ_0 of Γ_1 by some element of H. Thus H is k-homogeneous. $\square$

Note that the properties mentioned above, namely, k-transitivity, k-primitivity and k-homogeneity are *local* properties in the sense that a permutation group H on a set Σ has one of these properties $\mathcal{P}$ if and only if any finite subset Φ of Σ is contained in some subset Σ_0

for which there exists a subgroup H_0 of H stabilising Σ_0 setwise and inducing on it a permutation group that has property $\mathcal{P}$. The reader is encouraged to prove for himself the localness of these properties before proceeding any further. Using localness, the following corollary is a routine inference from the last lemma.

Corollary 10.10 *The union of a connected family of k-transitive (k-primitive) Jordan sets is a k-transitive (k-primitive) Jordan set.*

The union of a connected family of k-homogeneous Jordan sets, each of size $\geq 2k$, is a k-homogeneous Jordan set. $\square$

Theorem 10.11 *If G is primitive on Ω and there is a k-transitive (k-primitive) Jordan set then G is k-transitive (k-primitive) on Ω.*

If G is primitive on Ω and there is a k-homogeneous Jordan set of size $\geq 2k$ then G is k-homogeneous on Ω.

Proof: We prove the theorem for k-transitivity. The proofs for the other cases follow by similar arguments. Let $\alpha \in \Omega$ and let

$$\Omega_0 := \bigcup \{ \Gamma \mid \alpha \in \Gamma \text{ and } \Gamma \text{ is a } k\text{-transitive Jordan set} \}.$$

Then by the last corollary, Ω_0 is a k-transitive Jordan set. If $\Omega_0 \neq \Omega$, then since G is primitive, there exists $g \in G$ such that $\Omega_0^g \neq \Omega_0$ and $\Omega_0^g \cap \Omega_0 \neq \emptyset$. If $\Omega_0^g \subset \Omega_0$, we replace the element g by g^{-1}. In either case, we may assume that, $\Omega_0^g \not\subseteq \Omega_0$. Then $\Omega_0^g \cup \Omega_0$ is a k-transitive Jordan set containing α but strictly bigger than Ω_0, a contradiction. So $\Omega_0 = \Omega$. $\square$

Lemma 10.12 *Suppose that G acts transitively on Ω and that there is a Jordan set Γ which is maximal amongst proper subsets of Ω that are Jordan sets. Set $\Delta := \Omega \setminus \Gamma$. Then either Γ or Δ is a block of imprimitivity (possibly trivial) for G.*

Proof: Suppose that neither Γ nor Δ is a block. Then there exist $g, h \in G$ such that

$$\Gamma^g \neq \Gamma \quad \text{and} \quad \Gamma^g \cap \Gamma \neq \emptyset$$
$$\text{and } \Delta^h \neq \Delta \quad \text{and} \quad \Delta^h \cap \Delta \neq \emptyset.$$

If $\Gamma^g \subseteq \Gamma$ then $\Gamma^{g^{-1}}$ is a Jordan set containing Γ. So $\Gamma^g \not\subseteq \Gamma$. But then $\Gamma^g \cup \Gamma$ is a Jordan set properly containing Γ, and so we must have $\Gamma^g \cup \Gamma = \Omega$.

Now, $\Gamma^h \not\subseteq \cdot \Gamma$ for the same reason that $\Gamma^g \not\subseteq \Gamma$. So either

$$\Gamma^h \cap \Gamma = \emptyset \text{ or } \Gamma^h \cup \Gamma = \Omega.$$

But $\Delta^h \cap \Delta \neq \emptyset$ implies $\Gamma^h \cup \Gamma \neq \Omega$. So $\Gamma^h \cap \Gamma = \emptyset$ and therefore,

$$\Gamma^h \subseteq \Delta \subset \Gamma^g.$$

(The last containment follows from the fact that $\Gamma^g \cup \Gamma = \Omega$ with $\Gamma^g \cap \Gamma \neq \emptyset$.) But this means $\Gamma \subset \Gamma^{gh^{-1}}$, contradicting our choice of Γ. Hence either Γ or Δ is a block for G. $\square$

10.4   Cofinite Jordan sets

A *cofinite set* is a subset whose complement is finite.

Theorem 10.13 *Suppose that G is primitive on Ω and that there is a cofinite Jordan set $\Gamma \neq \Omega$. Then G is 2-transitive on Ω.*

Proof: Without loss of generality, we may suppose that the cofinite Jordan set $\Gamma \neq \Omega$ is chosen in such a way that $|\Delta|$ is minimal, where $\Delta := \Omega \setminus \Gamma$. Then Γ is maximal amongst proper subsets of Ω which are Jordan sets, and so by Lemma 10.12 either Γ or Δ is a block of imprimitivity. Since G is primitive, a proper subset of Ω which is a block of imprimitivity must be a singleton. But Γ, being a Jordan set, is not a singleton. Therefore $|\Delta| = 1$ and it follows immediately (see the case $k = 1$ of Thm. 3.13) that G is 2-transitive. $\square$

Theorem 10.14 *Suppose that G is primitive on Ω. Then a cofinite primitive Jordan set is improper.*

Proof: Let Γ_0 be a primitive cofinite Jordan set. If $k := |\Omega \setminus \Gamma_0|$, we may assume that $k > 0$. Let Γ_1 be a Jordan set minimally properly containing Γ_0. Such a set exists because Γ_0 is cofinite and Ω is a Jordan set properly containing Γ_0. Let $\Delta_0 := \Gamma_1 \setminus \Gamma_0$. We sometimes call Δ_0 a *minimal increment* for Γ_0. Then by Lemma 10.12 applied to the setwise stabiliser $G_{\{\Gamma_1\}}$, we have that either Δ_0 or Γ_0 is a block for $G_{\{\Gamma_1\}}$. If Ω is infinite then, since Γ_0 is cofinite, it cannot be a block of Γ_1. If Ω is finite, then by primitivity of G there exists $g \in G$ such that $\Gamma_0^g \neq \Gamma_0$ and $\Gamma_0^g \cap \Gamma_0 \neq \emptyset$; then $\Gamma_0^g \cup \Gamma_0$ is a Jordan set properly containing Γ_0 and it follows from the minimality of Γ_1 that $|\Gamma_1| \leq |\Gamma_0^g \cup \Gamma_0| < 2|\Gamma_0|$, so that again Γ_0 cannot be a block in Γ_1. Therefore Δ_0 is a block for $G_{\{\Gamma_1\}}$ in Γ_1.

Let ρ_0 be the $G_{\{\Gamma_1\}}$-congruence on Γ_1 for which Δ_0 is a class. Then $\rho_0|_{\Gamma_0}$ is a $G_{\{\Gamma_1\}\{\Delta_0\}}$-congruence. Hence it is also a $G_{(\Omega \backslash \Gamma_0)}$-congruence on Γ_0. Since Γ_0 is a primitive Jordan set, and Γ_0 is not a ρ_0-class, ρ_0 must be the trivial congruence, and this implies that Δ_0 is a singleton set.

From this it follows that Γ_1 is a 2-transitive cofinite Jordan set. Being 2-transitive, it is primitive. Hence we can now consider Γ_1 instead of Γ_0 and proceed as before. Continuing in this manner, we shall find primitive cofinite Jordan sets $\Gamma_0 \subset \Gamma_1 \subset \ldots \subset \Omega$ where each set consists of exactly one more element than the previous set. Since Γ_k will be $(k+1)$-transitive, and since on the other hand $\Gamma_k = \Omega$, the group G is also $(k+1)$-transitive. This means that Γ_0 is improper. $\square$

Note: The cases of Theorems 10.13 and 10.14 in which Ω is finite were proved by Camille Jordan in 1871.

Theorem 10.15 (Adeleke & Neumann 1996a) *Suppose that G is primitive on Ω. If there is a 3-transitive proper Jordan set Γ then G is highly transitive on Ω.*

Proof: By Lemma 10.11, the group G is 3-transitive. Let us assume, as inductive hypothesis, that G is k-transitive. By Theorem 10.14, the set Γ is coinfinite (that is, a set with an infinite complement). So we can choose $k - 2$ distinct points $\alpha_1, \alpha_2, \ldots, \alpha_{k-2}$ in $\Omega \backslash \Gamma$. But then their pointwise stabiliser $G_{\alpha_1, \alpha_2, \ldots, \alpha_{k-2}}$ is 2-transitive, and hence primitive, on $\Omega \backslash \{\alpha_1, \alpha_2, \ldots, \alpha_{k-2}\}$. Then by Lemma 10.11 again and the fact that Γ is 3-transitive, we must have $G_{\alpha_1, \alpha_2, \ldots, \alpha_{k-2}}$ is 3-transitive on $\Omega \backslash \{\alpha_1, \alpha_2, \ldots, \alpha_{k-2}\}$. But this implies that G is $(k+1)$-transitive on Ω. This completes the induction. $\square$

Note that a 2-primitive proper Jordan set suffices in the above theorem. We only need to apply Lemma 10.11 for k-primitive Jordan sets instead of k-transitive sets to prove the theorem.

We end this chapter by stating (without proof) a classification theorem of primitive groups with cofinite Jordan sets. The proof is by using induction on a geometry (in the sense of Definition 15.3) associated with such Jordan groups.

Theorem 10.16 (Neumann 1985) *Suppose that G is a primitive permutation group with the property that for every natural number n there exist cofinite Jordan sets whose complements have cardinality at least n. Then either G is highly transitive or it is a subgroup of a projective or an affine group of semilinear transformations over a finite field.* $\square$

Note that G is highly transitive if and only if every cofinite set is an (improper) Jordan set.

<u>EXERCISES:</u>

10(i) Suppose that groups A, B act regularly on sets Ω_1, Ω_2 respectively. Describe all Jordan sets for the wreath product $A \operatorname{Wr} B$ acting on $\Omega_1 \times \Omega_2$.

10(ii) Describe all Jordan sets for $\operatorname{Sym}(\Omega_1) \operatorname{Wr} \operatorname{Sym}(\Omega_2)$ acting on $\Omega_1 \times \Omega_2$.

10(iii) Let $(\Omega, \leq)$ be a partially ordered set. Show that if Γ is a Jordan set for $\operatorname{Aut}(\Omega, \leq)$ then Γ is convex (that is, if $\gamma_1, \gamma_2 \in \Gamma$ and $\gamma \in \Omega$ is such that $\gamma_1 \leq \gamma \leq \gamma_2$ then $\gamma \in \Gamma$).

10(iv) Let X be a 0-dimensional Hausdorff topological space (that is, a Hausdorff space in which the family of sets that are both open and closed forms a base for the topology). Let G be the group of all homeomorphisms $X \to X$ and suppose that G is transitive. Prove that a non-empty subset is a Jordan set for G if and only if it is open. In particular, show that a non-empty subset of $\mathbb{Q}$ is a Jordan set for the group of topological automorphisms of $\mathbb{Q}$ if and only if it is open.

10(v) Let Γ be a Jordan set for the transitive permutation group G on a set Ω, and let ρ_Γ be the G-congruence generated by all pairs (γ_1, γ_2) with $\gamma_1, \gamma_2 \in \Gamma$. Let ρ be any G-congruence. Prove that either $\rho < \rho_\Gamma$ or $\rho_\Gamma \leq \rho$, and that if $\rho < \rho_\Gamma$ then Γ is a union of ρ-classes.

10(vi) Complete the proof of Lemma 10.5(ii).

10(vii) Give examples to show how Lemma 10.7 fails for imprimitive groups.

10(viii) Suppose that Γ_1, Γ_2 are primitive Jordan sets for G in Ω and that they form a *typical pair*, (that is, $\Gamma_1 \cap \Gamma_2 \neq \emptyset$, $\Gamma_1 \not\subseteq \Gamma_2$ and $\Gamma_2 \not\subseteq \Gamma_1$). Prove that $\Gamma_1 \cup \Gamma_2$ is a 2-homogeneous Jordan set.

10(ix) Suppose that Γ_1, Γ_2 is a typical pair (as defined in the last exercise) of k-transitive Jordan sets for G in Ω. Prove that if $k \geq 2$ then $\Gamma_1 \cup \Gamma_2$ is a $(k+1)$-transitive Jordan set. Prove further that if either $\Gamma_1 \setminus \Gamma_2$ or $\Gamma_2 \setminus \Gamma_1$ is infinite then $\Gamma_1 \cup \Gamma_2$ is a highly transitive Jordan set.

10(x) Let G be a k-transitive Jordan group on the set Ω. Define a set Λ to be a *block* if there exist distinct points $\alpha_1, \ldots, \alpha_k \in \Lambda$ such that $\Omega \setminus \Lambda$ is a Jordan set that is maximal subject to containing none of $\alpha_1, \ldots, \alpha_k$. Show that the family of blocks forms a Steiner k-system on Ω (as defined in Defn. 11.2).

Chapter 11

Examples of Jordan Groups

We have already seen some examples of Jordan groups in the last chapter. We shall see many more examples in this chapter and the next. As the ultimate purpose of this study is to find a complete classification (in Chapter 13) of all infinite primitive Jordan groups which are not highly transitive, we shall discuss some of the other examples, namely the finite and highly transitive Jordan groups, quite briefly, in the next section. Thereafter, we shall concentrate on the classes of examples arising from the projective and affine spaces as well as from linear relational structures.

11.1 Some examples

In this section we shall state some examples of finite and highly transitive Jordan groups. Examples of imprimitive Jordan groups arise most naturally, as we have already mentioned in the last chapter, from constructions that look like wreath products.

11.1.1 Finite primitive examples

The easiest examples of finite groups with Jordan sets are the symmetric groups S_n and the alternating groups A_n. Sets of size ≥ 2 are Jordan sets for S_n while those of size ≥ 3 are Jordan sets for A_n. They are naturally primitive. But they are not Jordan groups as they are (sharply) n-transitive and $(n-2)$-transitive respectively.

A class of finite Jordan groups are those arising from $PGL(V)$ and $AGL(V)$ with V a finite dimensional vector space over a finite field F. The Jordan sets in $PGL(V)$ and $AGL(V)$ are the complements of·projective and affine subspaces in the corresponding spaces on which they act (cf. Lem. 11.1).

The alternating group A_7 is a Jordan group on the 15 points of the projective geometry $PG(3,2)$ and has Jordan sets of size 12. The group A_7 has a transitive extension (as defined in p. 142) by the translation group in the affine group $AGL(4,2)$. This extension is triply transitive and has Jordan sets of size 12.

The finite Mathieu groups M_{22}, M_{23}, M_{24} (cf. Gorenstein 1982) and $\text{Aut}(M_{22})$ are also Jordan groups of degrees 22, 23, 24 and 22 respectively. They act on certain Steiner systems (which we shall define in the next section) and have Jordan sets of size 16. As shown in Neumann (1985) and also in Kantor (1985), these are all the examples of finite primitive groups with proper Jordan sets.

11.1.2   Highly transitive examples

If $\text{FS}(\Omega) \leq G \leq \text{Sym}(\Omega)$, where $\text{FS}(\Omega)$ is the set of all permutations on Ω with finite support, then all subsets Σ of Ω of size ≥ 2 are primitive Jordan sets for G, and they are proper if they are cofinite. Similarly subsets of size ≥ 3 are primitive Jordan sets for G with $\text{Alt}(\Omega) \leq G \leq \text{Sym}(\Omega)$, where $\text{Alt}(\Omega)$ is the finitary alternating group on Ω (cf. Sec. 6.1).

Before stating the next example, let us recall that a topological space is 0-*dimensional* if the sets that are both open and closed form a base for the topology and *homogeneous* if for any points x, y there is a homeomorphism mapping x to y. If Ω is a homogeneous 0-dimensional Hausdorff topological space and $G := \text{Homeo}(\Omega)$ is the group of homeomorphisms of Ω to itself, then a subset of size > 1 is a Jordan subset if and only if it is open. It is a highly transitive Jordan set.

Fix $d \geq 2$. Let Ω be a d-dimensional path-connected manifold without boundary, and define $G := \mathrm{Homeo}(\Omega)$. A subset Σ of size > 1 is a primitive Jordan subset if and only if it is open and connected, and in this case it is also highly transitive. For further details see Examples 7.4.2 and 7.4.3 of Adeleke & Neumann (1996a).

11.2 Linear groups and Steiner systems

Let V be a vector space of dimension $n > 2$ over a field F and let $\mathrm{GL}(V)$ be the group of all non-singular linear transformations from V to itself. Then $\mathrm{GL}(V)$ has a natural action on V (cf. Sec. 7.2). Let $W < V$ be a proper subspace of V. Then $V \setminus W$ can be seen easily to be a Jordan set for $\mathrm{GL}(V)$ (compare the proof of the following lemma). This, however, does not give us a Jordan group straightaway, as $\mathrm{GL}(V)$ is not transitive on V. But $\mathrm{GL}(V)$ is a Jordan group in its action on the non-zero vectors of V.

The group $\mathrm{PGL}(V)$ acts 2-transitively (and hence primitively) on the projective space $\mathrm{PG}(V)$, the space of 1-dimensional subspaces of V (cf. Lem. 7.7). It is also a Jordan group as we shall see in the following lemma.

Lemma 11.1 *The projective linear group* $\mathrm{PGL}(V)$ *acting on the projective space* $\mathrm{PG}(V)$ *is a Jordan group. The Jordan complements for this group are precisely the k-flats of* $\mathrm{PG}(V)$.

Proof: Given any k-flat W (as defined in Section 7.2) of the projective space $\mathrm{PG}(V)$, and 1-dimensional spaces U_1 and U_2, generated by u_1 and u_2 respectively, not in W, we need to find an element in $\mathrm{PGL}(V)$ fixing all of W and mapping U_1 to U_2. But given any basis of W we can always extend it to get two bases of V, one containing u_1 and the other containing u_2. Since $\mathrm{GL}(V)$ is transitive on the set of ordered bases of V (cf. Ex. 7(i)), we can find a unique element in $\mathrm{GL}(V)$ fixing every element of the chosen basis of W (and hence all of W) and mapping u_1 to u_2. So every k-flat is a Jordan complement.

Conversely, let $X \subseteq PG(V)$ be a Jordan set of $\mathrm{PG}(V)$ and let H be a subgroup of $\mathrm{PGL}(V)$ associated with X. Define $U := V \setminus X^*$ where $X^* := \{v \in V \mid \langle v \rangle \in X\}$. We want to show that U is a subspace of V. Certainly $\hat{0} \in U$. Suppose that $u_1, u_2 \in U$ and

$\alpha, \beta \in F$. Since H fixes $\langle u_1 \rangle$ and $\langle u_2 \rangle$ its preimage H^* in $GL(V)$ fixes the subspace $\langle u_1, u_2 \rangle$ setwise. We claim that H^* fixes every 1-dimensional subspace of $\langle u_1, u_2 \rangle$. Then $\alpha u_1 + \beta u_2 \in U$ and so U would be a subspace, as required.

To prove our claim suppose that, on the contrary, there is a 1-dimensional subspace of $\langle u_1, u_2 \rangle$ that is not fixed setwise by every element of H^*. Then we must have $X^* \subseteq \langle u_1, u_2 \rangle$ and so every 1-dimensional subspace not contained in $\langle u_1, u_2 \rangle$ is fixed setwise by H^*. Since $\dim(V) \geq 3$ we can choose u_3 linearly independent from u_1 and u_2. For any $u \in \langle u_1, u_2 \rangle$ the 1-dimensional subspaces $\langle u_3 \rangle$ and $\langle u_3 + u \rangle$ have to be fixed setwise by H^*, and hence also the 2-dimensional subspace $\langle u_3, u_3 + u \rangle$. Then the space $\langle u_1, u_2 \rangle \cap \langle u_3, u_3 + u \rangle$ is also fixed setwise by H^*. But this is $\langle u \rangle$. Thus H^* fixes all 1-dimensional subspaces setwise. This means that H is the trivial subgroup of $PGL(V)$ and that X would have to be a singleton, contradicting the fact that it is a Jordan set. $\square$

Similarly, the affine group $AGL(V)$ can also be shown to be a (2-transitive) Jordan group. These Jordan groups, namely projective and affine groups, are special cases of groups preserving certain Steiner systems, which we define below.

Definition 11.2 Let k be a natural number greater than 2. A *Steiner k-system* $(\Omega, \mathcal{B})$ consists of a set Ω of *points* and a set $\mathcal{B}$ of *blocks* (or *Steiner lines*) of the Steiner system where elements of $\mathcal{B}$ are subsets of Ω, all of the same size greater than k, satisfying the following axioms:

1. There is more than one block.

2. If $\alpha_1, \alpha_2, \ldots, \alpha_k$ are distinct points of Ω then there is a unique $\Gamma \in \mathcal{B}$ such that $\alpha_1, \alpha_2, \ldots, \alpha_k \in \Gamma$.

An example of a Steiner 2-system is a pair with $PG(V)$ as the set of points and the 2-dimensional subspaces of V (regarding such a subspace as a collection of 1-dimensional subspaces contained in it) as the blocks. This gives us a Steiner 2-system because there is more than one block (as $\dim(V) \geq 3$) and also because any two distinct 1-dimensional subspaces lie in a unique 2-dimensional subspace of V. Clearly, this Steiner system is preserved under the action of $PGL(V)$

on $PG(V)$. Similarly, there is a Steiner system on $AG(V)$, the affine space over a field F, invariant under $AGL(V)$ (see Ex. 11(ii) below).

As an exercise to convince ourselves that we can crudely build Steiner k-systems for arbitrary k, fix a natural number $k > 1$. Choose $k + 1$ distinct points in the first step. Then at each even step, add a block for every set of k distinct points that is not already contained in some block. So in the second step we put in $k + 1$ blocks, one for each set of k distinct points from the set of $k + 1$ points we had chosen in the first step. At each odd step, we add a new point to each existing block. At the end we will have a Steiner k-system with infinite blocks. It is called the *free* Steiner k-system with infinite blocks.

EXERCISES:

11(i) What can you say about the automorphism group of the free Steiner k-system constructed in the last paragraph?

11(ii) Show that in the affine case, the action of $AGL(V)$ preserves a Steiner 2-system if and only if the field has more than 2 elements. Furthermore, if the field has only 2 elements, then it preserves a Steiner 3-system.

11(iii) For some $k > 2$ let $(\Omega, \mathcal{B})$ be a Steiner k-system. For $\alpha \in \Omega$ define $\Omega' := \Omega \setminus \{\alpha\}$ and $\mathcal{B}' := \{\Gamma \setminus \{\alpha\} \mid \alpha \in \Gamma \in \mathcal{B}\}$. Then show that $(\Omega', \mathcal{B}')$ is a Steiner $(k-1)$-system.

11(iv) Show that a k-transitive group of automorphisms G of a Steiner k-system on Ω cannot have a *primitive* Jordan set. [Hint: Let Γ be a proper Jordan set of such a Jordan group and pick $\alpha_0, \alpha_1, \ldots, \alpha_{k-1} \in \Omega \setminus \Gamma$. Let $\alpha_k \in \Gamma$ and let L denote the block containing $\alpha_1, \alpha_2, \ldots, \alpha_k$. Show that $L \cap \Gamma$ is a proper subset of Γ and so is a block of imprimitivity of the group $G_{(\Omega \setminus \Gamma)}$ acting on Γ (cf. Macpherson 1994, Remark 3, p. 82).]

11(v) Suppose that G is k-homogeneous on the countably infinite set Ω, with k an integer greater than 1, and suppose that for any k-subset Δ of Ω, the group $G_{(\Delta)}$ has finitely many orbits on $\Omega \setminus \Delta$, with at least one finite orbit. Let Γ be the union of the finite orbits of $G_{(\Delta)}$ on Ω. Then show that $\{\Gamma^g \mid g \in G\}$ is the set of blocks of a G-invariant Steiner k-system on Ω with finite blocks. [This result is due to Cameron (see Macpherson 1985).]

There are other examples of Jordan groups which preserve Steiner systems. We shall say that a Jordan group is of *geometric type* if it preserves a Steiner system. We shall prove, in Chapter 15, that Hrushovski's construction gives us a k-transitive, not $(k+1)$-transitive Jordan group of geometric type.

11.3   Linear relational structures

We have already studied the set of rationals under the natural linear order on it in Chapter 9. But there are other relations, besides the linear order, which can be defined on $\mathbb{Q}$. We are interested in them mainly because their automorphism groups give us more examples of Jordan groups. We discuss them below and then axiomatise the relations for arbitrary sets. We shall use the symbols $\forall$ for 'for all', $\exists$ for 'there exists', $\vee$ for 'or', $\wedge$ for 'and' and $\neg$ for negation in the discussion that follows. Proofs of most of the theorems that follow can be found in Huntington (1935). Our discussion is based on Section 1.3 of Adeleke & Neumann (1996c).

Before we begin, it might be best to understand how a relation between elements of a set can be described set-theoretically. For example, the binary relation $<$ defined on the set $\mathbb{Q}$ can be described as a subset $<$ of $\mathbb{Q} \times \mathbb{Q}$, and is then a set of pairs from $\mathbb{Q}$, namely $< = \{(x,y) \mid x, y \in \mathbb{Q}, x < y\}$. But relations need not be just binary.

Definition 11.3 For an integer $n \geq 2$, an *n-ary relation* R on a set Ω is a subset of Ω^n. The integer n is called the *arity* of the relation R.

We say that $R(a_1, a_2, \ldots, a_n)$ or that the n-tuple $(a_1, a_2, \ldots, a_n)$ is *R-related* whenever $(a_1, a_2, \ldots, a_n) \in R$. A 2-ary relation is called a *binary* relation, a 3-ary relation is *ternary* and a 4-ary relation is called *quaternary*. More generally, a relation is said to be *finitary* if it is n-ary for some finite n. For more on the subject of relations, see Fraïssé (1986).

Definition 11.4 If $\mathcal{R}$ is a set of finitary relations on Ω then the group of permutations of Ω which preserve all the relations in $\mathcal{R}$ is denoted as $\text{Aut}(\Omega, \mathcal{R})$. That is, $g \in \text{Aut}(\Omega, \mathcal{R})$ if $g \in \text{Sym}(\Omega)$ and for $a_1, a_2, \ldots, a_n \in \Omega$, we have

$$R(a_1, a_2, \ldots, a_n) \Leftrightarrow R(a_1^g, a_2^g, \ldots, a_n^g) \quad \text{for all } R \in \mathcal{R}.$$

An n-ary relation R is said to be *non-trivial* if at least one n-tuple is R-related, and *proper* if not all n-tuples are R-related. A set Ω with one or more relations defined on it is called a *relational structure*. If $\mathcal{R}$ contains only one relation R then we write $\text{Aut}(\Omega, \mathcal{R})$ simply as $\text{Aut}(\Omega, R)$.

Clearly, $\text{Aut}(\Omega, \mathcal{R})$ is a subgroup of $\text{Sym}(\Omega)$. Conversely, if $G \leq \text{Sym}(\Omega)$ and if R_n is an orbit of G on Ω^n for some $n \in \mathbf{N}$, then R_n is an n-ary relation and it is easy to see that $G \leq \text{Aut}(\Omega, R_n)$. If we introduce a relation symbol for every G-orbit on n-tuples for all $n \in \mathbf{N}$, then G is *dense* in the automorphism group $\text{Aut}(\Omega, \{R_n\}_{n \in \mathbf{N}})$ of the resulting structure in the sense that it has the same orbits on finite ordered sets. (Note that terms like 'dense' and 'closed' used in this context also have a topological interpretation (see Cameron 1990, p. 27).) Thus, we see that all permutation groups can be described as dense subgroups of automorphism groups of relational structures. This is the basis of the connection between permutation group theory and relational structures.

Suppose a set Ω has a non-trivial n-ary relation R defined on it such that if R holds of an n-tuple then all the entries in that tuple are distinct, but R does not hold for all such tuples. Then the automorphism group $\text{Aut}(\Omega, R)$ cannot be n-transitive because no automorphism of Ω can map an n-set of Ω which is in R to another set which is not. We shall use this observation repeatedly in the rest of this section.

11.3.1 Linear order

We have already stated the axioms defining a strict *linear order* (or a *total order*) in Chapter 9. We state them again here in the non-strict form. Of course, the strict version can be defined in terms of the definition that follows, and vice versa. Really, condition a1 of Definition

11.5 should read $\forall \alpha (\alpha \leq \alpha)$, and likewise the universal quantifiers should be added at the beginning of a2 and a3. For simplicity, we have deleted all universal quantifiers in the following conditions, and we shall do so later in the chapter too. We shall also assume that α, β, γ, etc. are elements from the set under consideration (for instance, α, β, γ are elements of Λ in the definition below), without explicitly saying so.

Definition 11.5 (i) A pair $(\Lambda, \leq)$ is called a *partially ordered set* or simply a *poset*) if

a1 $(\alpha \leq \alpha)$ (reflexivity);

a2 $(\alpha \leq \beta \wedge \beta \leq \alpha \Rightarrow \alpha = \beta)$ (anti-symmetry);

a3 $(\alpha \leq \beta \wedge \beta \leq \gamma \Rightarrow \alpha \leq \gamma)$ (transitivity).

If $\alpha \leq \beta$ but $\alpha \neq \beta$ we denote this as $\alpha < \beta$.

(ii) Two elements $\alpha, \beta \in \Lambda$ are said to be *comparable* if either $\alpha \leq \beta$ or $\beta \leq \alpha$. Otherwise they are said to be *incomparable*, and the fact is denoted by $\alpha \| \beta$. A partially ordered set Λ in which all elements are comparable is called a *linearly ordered set* or a *totally ordered set* or simply a *chain*.

The rationals have a natural linear ordering $\leq$ defined on them. We have already seen (in Chapter 9) that $\mathrm{Aut}\,(\mathbf{Q}, <)$, and hence also $\mathrm{Aut}\,(\mathbf{Q}, \leq)$, the group of all order-preserving automorphisms on $\mathbf{Q}$, is highly homogeneous. In particular, it is primitive and oligomorphic; but it is not 2-transitive. It is also a Jordan group, because any open interval in $\mathbf{Q}$ is a Jordan set. To see this, given $\alpha, \beta \in \mathbf{Q}$ and $I := (\alpha, \beta)$, let μ, ν be two points in I. Then it is an easy exercise (cf. Eg. 3(j)) to define order isomorphisms between the intervals (α, μ) and (α, ν), and between (μ, β) and (ν, β). Then there exists an order automorphism of $\mathbf{Q}$ extending these isomorphisms, taking μ to ν and fixing everything else in $\mathbf{Q}$ pointwise.

EXERCISE:

11(vi) Describe all Jordan sets of $\mathrm{Aut}\,(\mathbf{Q}, \leq)$.

11.3.2 Linear betweenness relation

A *linear betweenness relation* B on $\mathbf{Q}$ is a ternary relation defined (in terms of the natural linear order $\leq$ on $\mathbf{Q}$) by

$$B(\alpha; \beta, \gamma) :\Leftrightarrow (\beta \leq \alpha \leq \gamma) \vee (\gamma \leq \alpha \leq \beta).$$

In words, $B(\alpha; \beta, \gamma)$ if α lies in between β and γ under the natural ordering on $\mathbf{Q}$ (see Fig. 3).

$$\beta \qquad \alpha \ \gamma \qquad\qquad\qquad \gamma \quad \alpha \qquad \beta$$
$$\beta \leq \alpha \leq \gamma \qquad\qquad\qquad \gamma \leq \alpha \leq \beta$$

Figure 3: $B(\alpha; \beta, \gamma)$

Consider $\mathrm{Aut}\,(\mathbf{Q}, B)$, the group of all automorphisms of $\mathbf{Q}$ that preserve the relation B. Clearly, any element of $\mathrm{Aut}\,(\mathbf{Q}, \leq)$ preserves B. Therefore, $\mathrm{Aut}\,(\mathbf{Q}, \leq) \leq \mathrm{Aut}\,(\mathbf{Q}, B)$. Also the involutory map $g : x \mapsto -x$ for all $x \in \mathbf{Q}$ is an automorphism of $\mathbf{Q}$ preserving B but reversing $\leq$. Therefore, $\mathrm{Aut}\,(\mathbf{Q}, \leq) < \mathrm{Aut}\,(\mathbf{Q}, B)$. In fact, it is not very difficult to see that

$$\mathrm{Aut}\,(\mathbf{Q}, B) = \langle \mathrm{Aut}\,(\mathbf{Q}, \leq), g \rangle \text{ and } |\mathrm{Aut}\,(\mathbf{Q}, B) : \mathrm{Aut}\,(\mathbf{Q}, \leq)| = 2.$$

Each Jordan set of $\mathrm{Aut}\,(\mathbf{Q}, \leq)$ is also a Jordan set of $\mathrm{Aut}\,(\mathbf{Q}, B)$. So in particular, open intervals in $\mathbf{Q}$ are Jordan sets. Also since $\mathrm{Aut}\,(\mathbf{Q}, \leq)$ is highly homogeneous (cf. Thm. 9.5), so is $\mathrm{Aut}\,(\mathbf{Q}, B)$. We have also seen that $\mathrm{Aut}\,(\mathbf{Q}, \leq)$ is not 2-transitive. But $\mathrm{Aut}\,(\mathbf{Q}, B)$ is 2-transitive, because the stabiliser of a point, say 0, in $\mathrm{Aut}\,(\mathbf{Q}, B)$ is transitive on $\mathbf{Q} \setminus \{0\}$. (Clearly, it is transitive on each of the open intervals $(-\infty, 0)$ and $(0, \infty)$ and the automorphism g fixes 0 and maps one to the other.) $\mathrm{Aut}\,(\mathbf{Q}, B)$ is not however, 2-primitive as the stabiliser of 0 in it has two blocks $(-\infty, 0)$ and $(0, \infty)$ and is therefore not primitive. It cannot be 3-transitive as it is not 2-primitive.

We axiomatise the linear betweenness relations for arbitrary sets in the following definition.

Definition 11.6 A ternary relation B defined on a set Λ is said to be a *linear betweenness relation* if the following hold:

b1 $B(\alpha; \beta, \gamma) \Rightarrow B(\alpha; \gamma, \beta)$;

b2 $B(\alpha; \beta, \gamma) \wedge B(\beta; \alpha, \gamma) \Leftrightarrow \alpha = \beta$;

b3 $B(\alpha; \beta, \gamma) \Rightarrow B(\alpha; \beta, \delta) \vee B(\alpha; \gamma, \delta)$;

b4 $B(\alpha; \beta, \gamma) \vee B(\beta; \gamma, \alpha) \vee B(\gamma; \alpha, \beta)$.

We have already seen, as in the case of the rationals, how a ternary relation can be derived from a linear order. That it is a linear betweenness relation is a consequence of the next theorem.

Theorem 11.7 *If $(\Lambda, \leq)$ is a linearly ordered set, then the ternary relation B derived from it by the rule*

$$B(\alpha; \beta, \gamma) :\Leftrightarrow (\beta \leq \alpha \leq \gamma) \vee (\gamma \leq \alpha \leq \beta)$$

is a linear betweenness relation.

Conversely, if B is a linear betweenness relation on a set Λ then there are precisely two linear order relations on Λ of which B is the derived betweenness relation, and one is the reverse of the other. $\square$

11.3.3   Circular order

If we twist the linear order on $\mathbf{Q}$ around at the two ends we obtain a circular order (see Fig. 4).

A *circular order* on $\mathbf{Q}$ is a ternary relation $\mathcal{K}$ defined on $\mathbf{Q}$ by

$$\mathcal{K}(\alpha, \beta, \gamma) :\Leftrightarrow (\alpha \leq \beta \leq \gamma) \vee (\gamma \leq \alpha \leq \beta) \vee (\beta \leq \gamma \leq \alpha).$$

Clearly, we have

$$\mathcal{K}(\alpha, \beta, \gamma) \Leftrightarrow \mathcal{K}(\beta, \gamma, \alpha) \Leftrightarrow \mathcal{K}(\gamma, \alpha, \beta),$$

and when α, β and γ are distinct, we have

$$\mathcal{K}(\alpha, \beta, \gamma) \Leftrightarrow \neg\mathcal{K}(\beta, \alpha, \gamma).$$

The automorphism group $\text{Aut}(\mathbb{Q}, \mathcal{K})$ of the rationals preserving $\mathcal{K}$ is 2-transitive and 2-primitive but not 3-transitive. It is a Jordan group because $\text{Aut}(\mathbb{Q}, \leq) \leq \text{Aut}(\mathbb{Q}, \mathcal{K})$. An automorphism preserving $\mathcal{K}$ need not preserve $\mathcal{B}$. Also, the involution map $g \in \text{Aut}(\mathbb{Q}, \mathcal{B})$ is not in $\text{Aut}(\mathbb{Q}, \mathcal{K})$. Clearly, $\text{Aut}(\mathbb{Q}, \mathcal{K})$ is transitive on $\mathbb{Q}$. When we remove a point, say 0, from $\mathbb{Q}$, the resulting linear ordering on $\mathbb{Q} \setminus \{0\}$ (see Thm. 11.9) is (either as defined from $\mathcal{K}$, or directly defined from the usual order on $\mathbb{Q}$ and therefore is) a dense linear order without endpoints and is hence transitive (but not 2-transitive) and primitive. Thus $\text{Aut}(\mathbb{Q}, \mathcal{K})$ is 2-transitive and 2-primitive. It is not 3-transitive.

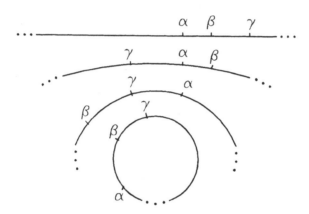

Figure 4: $\mathcal{K}(\alpha, \beta, \gamma)$

We now give the axioms for a circular order in general.

Definition 11.8 A *circular* (or *cyclic*) *order* is a ternary relation $\mathcal{K}$ defined on a set Λ satisfying the following conditions:

c1 $\quad \mathcal{K}(\alpha, \beta, \gamma) \Rightarrow \mathcal{K}(\beta, \gamma, \alpha)$;

c2 $\quad \mathcal{K}(\alpha, \beta, \gamma) \wedge \mathcal{K}(\beta, \alpha, \gamma) \Leftrightarrow \alpha = \beta \vee \beta = \gamma \vee \gamma = \alpha$;

c3 $\quad \mathcal{K}(\alpha, \beta, \gamma) \Rightarrow \mathcal{K}(\alpha, \beta, \delta) \vee \mathcal{K}(\delta, \beta, \gamma)$;

c4 $\quad \mathcal{K}(\alpha, \beta, \gamma) \vee \mathcal{K}(\beta, \alpha, \gamma)$.

The circular order is said to be *dense* if, for all distinct $\alpha, \beta \in \Lambda$, there is $\gamma \in \Lambda$ with $\mathcal{K}(\alpha, \beta, \gamma)$.

Using the same rule as for the rationals, a ternary relation $\mathcal{K}$ can be derived from a binary relation $\leq$. The following theorem relates linear orders and circular orders.

Theorem 11.9 *If $(\Lambda, \leq)$ is a linearly ordered set and $\mathcal{K}$ is the ternary relation derived from $\leq$ by the rule*

$$\mathcal{K}(\alpha, \beta, \gamma) :\Leftrightarrow (\alpha \leq \beta \leq \gamma) \vee (\gamma \leq \alpha \leq \beta) \vee (\beta \leq \gamma \leq \alpha)$$

then $\mathcal{K}$ is a circular order relation on Λ.

Conversely, if $(\Omega, \mathcal{K})$ is a circular ordering and $\alpha \in \Omega$, then the relation $\leq$ defined on $\Lambda := \Omega \setminus \{\alpha\}$ by the rule

$$\beta \leq \gamma :\Leftrightarrow \mathcal{K}(\alpha, \beta, \gamma)$$

is a linear order.

Furthermore, if we extend this order relation to one on Ω by specifying that $\beta < \alpha$ for all $\beta \in \Lambda$ then the derived circular order relation is the original circular order $\mathcal{K}$. □

Cyclic order relations always come in pairs. If $(\Omega, \mathcal{K})$ is a circularly ordered set and we define $\mathcal{K}^*$ by $\mathcal{K}^*(\alpha, \beta, \gamma) :\Leftrightarrow \mathcal{K}(\alpha, \gamma, \beta)$ then $\mathcal{K}^*$ can be easily verified to be a circular order relation on Ω, called the *reverse* of the relation $\mathcal{K}$.

11.3.4 Separation relation

There is a group of permutations of a circular order which preserve or reverse it, much as $\mathrm{Aut}\,(\mathbf{Q}, \mathcal{B})$ consists of permutations preserving or reversing a linear order. This gives rise to a quaternary separation relation $\mathcal{S}$ defined on $\mathbf{Q}$ by

$$\mathcal{S}(\alpha, \beta; \gamma, \delta) :\Leftrightarrow (\mathcal{K}(\alpha, \beta, \gamma) \wedge \mathcal{K}(\alpha, \delta, \beta)) \vee (\mathcal{K}(\alpha, \gamma, \beta) \wedge \mathcal{K}(\alpha, \beta, \delta)).$$

In other words, $\mathcal{K}(\alpha, \beta; \gamma, \delta)$ holds if and only if γ and δ lie on different 'components' of $\mathbb{Q} \setminus \{\alpha, \beta\}$ in the circular order (see Fig. 5).

Figure 5: $\mathcal{S}(\alpha, \beta; \gamma, \delta)$

The automorphism group $\text{Aut}\,(\mathbb{Q}, \mathcal{S})$ is a 3-transitive, not 3-primitive Jordan group. It contains each of $\text{Aut}\,(\mathbb{Q}, \leq)$, $\text{Aut}\,(\mathbb{Q}, \mathcal{B})$ and $\text{Aut}\,(\mathbb{Q}, \mathcal{K})$ (and is, in fact, generated by them). Fix $\alpha \in \Omega$, and define a relation $\mathcal{B}$ on $\Omega \setminus \{\alpha\}$, putting $\mathcal{B}(\beta; \gamma, \delta)$ if and only if $\mathcal{S}(\alpha, \beta; \gamma, \delta)$. Then an automorphism in $\text{Aut}\,(\mathbb{Q}, \mathcal{S})$ stabilising α is just an automorphism of the betweenness relation $\mathcal{B}$ such that $\mathcal{B}(\beta; \gamma, \delta)$. Conversely, any automorphism of this betweenness relation lifts to an automorphism of $\text{Aut}\,(\mathbb{Q}, \mathcal{S})$ fixing α. Since $\text{Aut}\,(\mathbb{Q} \setminus \{\alpha\}, \mathcal{B})$ is 2-transitive (and therefore primitive), it follows that $\text{Aut}\,(\mathbb{Q}, \mathcal{S})$ is 3-transitive and 2-primitive. It is not 3-primitive because $\text{Aut}\,(\mathbb{Q} \setminus \{\alpha\}, \mathcal{B})$ is not 2-primitive. It is not 4-transitive because it is not 3-primitive.

Another way to see this is as follows. Let α and β be two distinct points of $\mathbb{Q}$. In the circular order, $\mathbb{Q} \setminus \{\alpha, \beta\}$ is composed of two dense linear orders without endpoints. Also since $\mathcal{S}(\alpha, \beta; \gamma, \delta) \Leftrightarrow \mathcal{S}(\alpha, \beta; \delta, \gamma)$ the stabiliser of α and β in $\text{Aut}\,(\mathbb{Q}, \mathcal{S})$ is transitive on $\mathbb{Q} \setminus \{\alpha, \beta\}$. Therefore, $\text{Aut}\,(\mathbb{Q}, \mathcal{S})$ is 3-transitive on $\mathbb{Q}$. However, the two components of $\mathbb{Q} \setminus \{\alpha, \beta\}$ form blocks of imprimitivity for this stabiliser and therefore $\text{Aut}\,(\mathbb{Q}, \mathcal{S})$ is not 3-primitive.

A separation relation, in general, is an abstraction of the geometrical idea of two pairs of points on a circle separating each other (see Fig. 5). We axiomatise it in the following way.

Definition 11.10 A quaternary relation S defined on a set Λ is a *separation relation* if it satisfies the following conditions:

s1 $S(\alpha,\beta;\gamma,\delta) \Rightarrow S(\beta,\alpha;\gamma,\delta) \wedge S(\gamma,\delta;\alpha,\beta)$;

s2 $S(\alpha,\beta;\gamma,\delta) \wedge S(\alpha,\gamma;\beta,\delta) \Leftrightarrow \beta = \gamma \vee \alpha = \delta$;

s3 $S(\alpha,\beta;\gamma,\delta) \Rightarrow S(\alpha,\beta;\gamma,\epsilon) \vee S(\alpha,\beta;\delta,\epsilon)$;

s4 $S(\alpha,\beta;\gamma,\delta) \vee S(\alpha,\gamma;\delta,\beta) \vee S(\alpha,\delta;\beta,\gamma)$.

The following theorem gives the relation between separation relations and circular orders, and between separation relations and betweenness relations.

Theorem 11.11 (i) *If* $(\Omega,\mathcal{K})$ *is a circular ordering, then the quaternary relation* S *derived from it by the rule*

$$S(\alpha,\beta;\gamma,\delta) :\Leftrightarrow (\mathcal{K}(\alpha,\beta,\gamma) \wedge \mathcal{K}(\alpha,\delta,\beta)) \vee (\mathcal{K}(\alpha,\gamma,\beta) \wedge \mathcal{K}(\alpha,\beta,\delta))$$

is a separation relation on Ω.

Conversely, if S *is a separation relation on a set* Ω *then there are precisely two circular order relations on* Ω *of which* S *is the derived separation relation, and each is the reverse of the other.*

(ii) *If* B *is a linear betweenness relation on* Ω *and* S *is a quaternary relation derived from* B *by the rule*

$$S(\alpha,\beta;\gamma,\delta) :\Leftrightarrow (B(\alpha;\beta,\gamma) \wedge B(\delta;\alpha,\beta)) \vee$$
$$(B(\beta;\gamma,\alpha) \wedge B(\delta;\alpha,\beta)) \vee (B(\gamma;\alpha,\beta) \wedge \neg B(\delta;\alpha,\beta))$$

then S *is a separation relation on* Ω.

In the other direction, let S *be a separation relation on* Ω, $\alpha \in \Omega$ *and* $\Lambda := \Omega \setminus \{\alpha\}$. *If* B *is defined on* Λ *by the rule*

$$B(\beta;\gamma,\delta) :\Leftrightarrow S(\alpha,\beta;\gamma,\delta)$$

then B *is a linear betweenness relation.* $\square$

Replacing $\mathbb{Q}$ by any set with a linear ordering, we can similarly define each of the relations that we just defined on $\mathbb{Q}$ on the set. Since the automorphism groups contain the automorphism group of the underlying linear order, the automorphism groups of these relational structures form Jordan groups, provided the linear ordering is 2-homogeneous. Furthermore, if the linear ordering is 2-homogeneous and of size > 2 then all these groups are also highly homogeneous. But they are not highly transitive as they preserve certain relations. The following beautiful and fundamental result classifies these groups.

Theorem 11.12 (Cameron 1976) *Let G be a permutation group on an infinite set Ω, and suppose that G is highly homogeneous but not highly transitive. Then G preserves either a linear order or a circular order or a linear betweenness relation or a separation relation.* □

Note that each of the four relations that we get from the last theorem is dense, in the natural sense.

EXERCISES:

11(vii) Define a natural cyclic order on the roots of unity (which is a countable set) in the complex plane and show that it is dense.

11(viii) Show that the group of cyclic-order preserving permutations of $\mathbb{Q}$ is simple. [Hint: Any point stabiliser of this group is a permutation group isomorphic to $\mathrm{Aut}\,(\mathbb{Q}, <)$, so the normal subgroups of a point stabiliser are known (cf. Thm. 9.9).]

11(ix) (McDermott) Any 3-homogeneous not 2-transitive group preserves a linear order (cf. Cameron 1990, Sec.3.11). [Hint: Show that there is an invariant tournament (cf. p. 153) and look at possible structures of 3-element subtournaments. If none of these are cycles, show that there is an invariant linear order.]

11(x) Define a separation relation on $\mathrm{PG}(1, \mathbb{Q})$ using stereographic projection such that it is preserved by $\mathrm{PGL}(2, \mathbb{Q})$.

Chapter 12

Relations related to betweenness

In the last chapter we defined linear betweenness relations, circular (or cyclic) orders and separation relations from a linear order and studied their groups of automorphisms. The automorphism group of a linear order has already been studied in detail in Chapter 9. That of a circular order can be understood best in terms of the linear order obtained by deleting a point. The groups of automorphisms of a linear betweenness relation and of a separation relation are simply the groups of order-preserving or order-reversing transformations on a linearly ordered and cyclically ordered set respectively. In this chapter we introduce four more relations related to betweenness which will lead us to the classification of primitive Jordan groups that have primitive Jordan sets. Everything discussed in this chapter has been discussed in greater detail and rigour in Adeleke & Neumann (1996c). The arguments used in this chapter are very geometric and we encourage the reader to draw pictures. Note however that our semilinear orders grow upwards rather than downwards, contrary to the convention followed in Adeleke & Neumann (1996c).

12.1 Semilinear order

A *semilinear order* is a partially ordered set in which for every point, the set of all points below it is linearly ordered and is 'connected' in a natural sense (see Fig. 6).

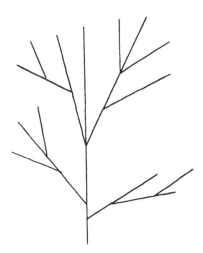

Figure 6: A Semilinear Order

We axiomatise the definition of a semilinear order below. As in the last chapter, we shall not write the universal quantifiers.

Definition 12.1 A partially ordered set $(P, \leq)$ is a *(lower) semilinearly ordered set* or simply a *(lower) semilinear order* if

A1 $(y \leq x \wedge z \leq x) \Rightarrow (y \leq z \vee z \leq y)$; in other words, for every $a \in P$ the set $\{x \in P \,|.x \leq a\}$ is totally ordered;

A2 $(\exists z)(z \leq x \wedge z \leq y)$;

A3 $(P, \leq)$ is not a total order.

As in Chapter 9 we say that $(P, \leq)$ is *unbounded* or *without endpoints* if it contains no maximal or minimal element. The set $(P, \leq)$ is said to be *dense* if $x < y \Rightarrow (\exists z)(x < z < y)$. Reversing the ordering, (upper) semilinearly ordered sets can be defined similarly.

Let us first convince ourselves of the existence of semilinearly ordered sets by constructing some examples.

12.2 Constructing semilinear orders:

We shall build a semilinear order, which we denote as $P(\mathbf{Q}, 2, +)$, from the linear order on the rationals. We begin with a copy of the rational line, say $\mathbf{Q}_0$. At the first stage, we adjoin a copy of $\mathbf{Q} \cap (q_i, \infty)$ at each point q_i of $\mathbf{Q}_0$. Call these the $\mathbf{Q}_1$ branches. The copy adjoined at the point q_0 will carry its natural ordering, and its elements will be greater than all elements of $\{x \in \mathbf{Q}_0 : x \leq q_0\}$, and will be incomparable to the rest of $\mathbf{Q}_0$ and to all other elements adjoined at the first stage. At the second stage, we do the same thing at each point of each of the $\mathbf{Q}_1$ branches. We keep on adjoining copies of parts of the rational line infinitely many times in infinitely many stages to finally obtain our structure $P := P(\mathbf{Q}, 2, +)$. The dense linear order on $\mathbf{Q}$ induces a natural dense lower semilinear order on P. It is clear, from the method of construction, that all maximal chains in P are isomorphic to $(\mathbf{Q}, \leq)$. (Note that we could have started with a copy of the natural numbers instead of the rationals and built a semilinear order $P(\mathbf{N}, 2, +)$ by a similar process. That order however, would not have been dense.)

Using the fact that $\mathrm{Aut}\,(\mathbf{Q}, \leq)$ is 2-homogeneous one can show that the automorphism group $\mathrm{Aut}\,(P(\mathbf{Q}, 2, +))$ is transitive on P. For a proof we refer the reader to Section 6 of Droste (1985) (specifically to Corollary 6.9 therein), although the notation there is somewhat different. $\mathrm{Aut}\,(P(\mathbf{Q}, 2, +))$ is also primitive (cf. Adeleke & Neumann 1996c, Thm. 6.1). It is not 2-transitive, simply because for example, distinct elements say $x, y \in P$ with $x \leq y$ cannot be mapped to y, x. But $\mathrm{Aut}\,(P(\mathbf{Q}, 2, +))$ is transitive on the set of unordered pairs of comparable elements in P.

This partial order has the property that any isomorphism between subsets of size at most 2 extends to an automorphism. This property is called *relative 2-transitivity* or *order 2-transitivity* in Adeleke & Neumann (1996c). Relatively k-transitive sets can be defined analogously. The property that every isomorphism between subsets of size exactly k extends to an automorphism is called *k-homogeneity* by Droste (1985). Note that this definition of homogeneity is quite distinct from the notion of homogeneity for permutation groups (and referred to, for example, in the beginning of the last paragraph). There is a classification of countable 2-homogeneous semilinear orders in Droste (1985).

12.3 Alternative Construction:

The lower semilinear order $P(\mathbb{Q}, 2, +)$ constructed above can also be considered to be a collection of finite sequences of rationals as we shall see now. Define P' to be the set of all finite sequences

$$q_1 q_2 \cdots q_k$$

with $k \geq 1, q_i \in \mathbb{Q}$ and $q_1 < q_2 < \ldots < q_k$. Given two sequences

$$q = q_1 q_2 \ldots q_k \text{ and } q' = q_1' q_2' \ldots q_{k'}'$$

we say that $q \leq q'$ if and only if

(i) $k \leq k'$,

(ii) $q_i = q_i'$ for $1 \leq i \leq k - 1$, and

(iii) $q_k \leq q_k'$.

Note that this ordering is just a version of the familiar *lexicographic* order. Routine checking shows that this is a lower semilinear order on P'.

It is easy to see that $P(\mathbb{Q}, 2, +)$ and $(P', \leq)$ are indeed isomorphic. Given an element $q = q_1 q_2 \ldots q_k$ in P', we can find an element in P corresponding to it, such that it lies on the maximal chain which branches from $\mathbb{Q}_0$ at the point q_1, then continues along the $\mathbb{Q}_1$ branch till it reaches q_2, branches there and continues along the $\mathbb{Q}_2$ branch till it reaches q_3 and so on till it reaches a $\mathbb{Q}_{k-1}$ branch. We go along that branch till we reach q_k. Call that point p and let $q \in P'$ correspond to that point $p \in P$ (see Fig. 7). Conversely, by reversing this process, any point on P can be expressed as a sequence of rationals of the form described for elements of P'. Clearly, the order relation is the same in these two constructions.

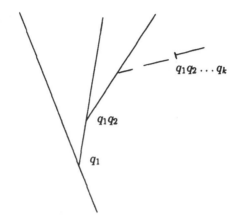

Figure 7: Alternative Construction

There is nothing special about adjoining a single copy of the rational line at each point of the previous stage, as we have done in the construction described in 12.2. We could as well have adjoined more than one copy at each point to get other examples of semilinearly ordered sets. Adjoining $n-1$ copies at each point would give us the semilinear order $P(\mathbb{Q}, n, +)$.

For a different kind of variation, we could have started with a coloured copy of the rationals coloured densely in two colours, red and green say, (cf. Sec. 9.2). We could then do the same construction as in 12.2, except that we put in coloured copies of the rational line, and only at the red points. After this is done, we discard all the red points. What we will be left with is still a semilinearly ordered set, except that, in this case, incomparable elements will never have a greatest lower bound. This object is said to be of *negative type* and denoted as $P(\mathbb{Q}, n, -)$ to distinguish this case from the previous case, called the *positive type*, in which the greatest lower bound of every pair of elements does exist. We shall restrict ourselves to semilinear orders of positive type in our discussion.

Let $P := (P, \leq)$ be a lower semilinear order, and let us fix $a \in P$. We sometimes refer to the elements of P as *nodes*. We can then partition the set P into three disjoint parts, one containing all elements $x > a$, the other $x \leq a$ and the third containing all the incomparable elements $x \| a$. Call the first set Γ. That is, $\Gamma := \{x \in P \mid a < x\}$. Clearly, automorphisms of Γ and of its complement can be pieced together to obtain automorphisms of the whole set P. This is called 'piecewise patching'. Consider an element $f \in \text{Sym}(P)$ that induces $g \in \text{Aut}(\Gamma, \leq)$ on Γ and the identity on $P \setminus \Gamma$. Then $f \in \text{Aut}(P, \leq)$. Thus if $\text{Aut}(P, \leq)$ and $\text{Aut}(\Gamma, \leq)$ are transitive then $\text{Aut}(P, \leq)$ is a Jordan group with Γ as a Jordan set. So semilinear orders are likely to give us examples of Jordan sets.

For any $a \in P$ and Γ as above, define a relation E_a on Γ by

$$x E_a y \Leftrightarrow (\exists z)(a < z \leq x \wedge a < z \leq y)$$

(see Fig. 8). This is easily seen to be an equivalence relation on Γ. We call the E_a-classes the *cones* at the node a. The equivalence relation E_a is invariant under the stabiliser of a in $\text{Aut}(P, \leq)$. The number of cones at each node is sometimes called the *ramification order*. Of course, the ramification order is well-defined on a semilinear order, only when this number is the same at each node. This is automatic, for example, when we have transitivity. The semilinear order $P(\mathbb{Q}, n, +)$, defined earlier, has ramification order n.

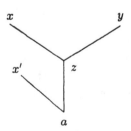

Figure 8: $x E_a y$ but $x' \not\!\!E_a y$

In $P(\mathbb{Q}, n, +)$, the stabiliser of a node a acts on Γ as the wreath product of the automorphism group of a cone at a by S_n where n is the ramification order. This shows that Γ is a Jordan set, as cones admit transitive automorphism groups. In the examples we have constructed, all the cones at a node a are isomorphic to each other and to $P(\mathbb{Q}, n, +)$. Hence the automorphism group of each cone is transitive and therefore, by an argument similar to one given earlier, is a Jordan set for $\mathrm{Aut}\,(P, \leq)$. Also, given any set S of cones at a node, any automorphism of $\bigcup S$, extended by the identity elsewhere, is an automorphism of $(P, \leq)$. Therefore, in $P(\mathbb{Q}, n, +)$, the union of any set of cones at a fixed node is a Jordan set. Likewise, for example by Lemma 10.5(ii), the union of any chain of cones, totally ordered by inclusion, is a Jordan set. In fact, primitive Jordan sets for $\mathrm{Aut}\,(P, \leq)$ will always be cones or unions of chains of cones. Thus $\mathrm{Aut}\,(P(\mathbb{Q}, n, +))$ is a primitive Jordan group. Theorem 6.9 of Adeleke & Neumann (1996c) gives a classification of Jordan sets for the semilinear orders we have seen in this section.

12.2   *C*-relations

Let Σ be the set of all maximal chains in a 2-homogeneous semilinear order $(P, \leq)$. For instance, in the terminology of the construction defined in 12.3, a maximal chain will be an infinite sequence

$$q_1 q_2 \cdots q_k \cdots$$

with $q_1 < q_2 < \ldots < q_k < \ldots$ where $q_i \in \mathbb{Q}$. Clearly, there are $2^{\aleph_0}$ such chains and $\mathrm{Aut}\,(P, \leq)$ acts on Σ too. There is a ternary relation C on Σ preserved by this action: for $\alpha, \beta, \gamma \in \Sigma$ we say that $C(\alpha; \beta, \gamma)$ if we have

$$(\beta = \gamma \neq \alpha) \vee (\alpha \cap \beta = \alpha \cap \gamma \subset \beta \cap \gamma),$$

regarding α, β, γ as subsets of P (see Fig. 9). This is the motivation behind the formal definition of a C-relation (C for 'chain') on an arbitrary set which we state below.

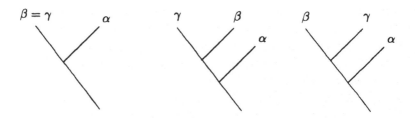

Figure 9: $C(\alpha; \beta, \gamma)$

Definition 12.4 Let Ω be a non-empty set and C a ternary relation on Ω. Then C is said to be a *C-relation* on Ω if

C1 $C(\alpha; \beta, \gamma) \Rightarrow C(\alpha; \gamma, \beta)$;

C2 $C(\alpha; \beta, \gamma) \Rightarrow \neg C(\beta; \alpha, \gamma)$;

C3 $C(\alpha; \beta, \gamma) \Rightarrow (C(\alpha; \delta, \gamma) \vee C(\delta; \beta, \gamma))$;

C4 $(\alpha \neq \beta) \Rightarrow C(\alpha; \beta, \beta)$;

C5 $(\exists \alpha) C(\alpha; \beta, \gamma)$;

C6 $(\alpha \neq \beta) \Rightarrow (\exists \gamma)(\beta \neq \gamma \wedge C(\alpha; \beta, \gamma))$.

It is easy to see that the relation defined earlier on the set Σ of maximal chains of a semilinear order satisfies all the above axioms and is therefore a C-relation. The following theorem says that this is essentially the only way to obtain C-relations.

Theorem 12.5 (Adeleke & Neumann 1996c, Theorem 12.4)
Let (Σ, C) be a C-relation. Then there is a semilinear order $(P, \leq)$ such that Σ is a set of maximal chains of $(P, \leq)$ with the relation C derived from it by the rule

$$C(\alpha; \beta, \gamma) :\Leftrightarrow \alpha \cap \beta = \alpha \cap \gamma \subset \beta \cap \gamma.$$

Also $P = \bigcup \Sigma$ in the sense that P is the union of all the internal nodes of the maximal chains in Σ.

Hint of proof: If (Σ, C) comes from a semilinear order $(P, \leq)$ (in which any two nodes have a greatest lower bound), then any pair of distinct chains determines a node, and any node is determined by a pair of chains; for, given $x \in P$, there are distinct $\beta, \gamma \in \Sigma$ such that $x = \max(\beta \cap \gamma)$ (and given distinct β, γ, the above x exists). (Here, we regard elements of Σ as sets of nodes.) But, of course, different pairs of chains may determine the same node. To prove the theorem we must recover P and its ordering from (Σ, C). So we define an equivalence relation $\approx$ on the set of unordered pairs of distinct elements of Σ, capturing the notion that two pairs determine the same node. We then define P to be the set of equivalence classes, and define a semilinear order on this set. We define $\approx$ by putting $\{\alpha, \beta\} \approx \{\gamma, \delta\}$ if and only if

$$\neg C(\alpha; \gamma, \delta) \wedge \neg C(\beta; \gamma, \delta) \wedge \neg C(\gamma; \alpha, \beta) \wedge \neg C(\delta; \alpha, \beta). \quad \Box$$

By the above result, it makes sense to refer to the underlying semilinear order $(P, \leq)$ of a C-relation, and to talk of elements of P as nodes, and to elements of Σ as chains.

12.6 Constructing C-relations:

Consider the set Σ of all sequences of finite support indexed by $\mathbb{Q}$ with entries in $\{0, 1\}$. Given $\sigma_1, \sigma_2, \sigma_3 \in \Sigma$, define $C(\sigma_1; \sigma_2, \sigma_3)$ if either $\sigma_2 = \sigma_3 \neq \sigma_1$ or if the least index (in $\mathbb{Q}$) at which σ_1 and σ_2 differ is the same and comes before the least index at which σ_2 and σ_3 differ. It is not difficult to show that this does define a C-relation on Σ.

Observe that the finite support condition in the construction above is only to ensure that the structure is countable. We can obtain a C-relation (on a set of size $2^{\aleph_0}$) even without the finite support condition. In the latter case (as well as in the instance cited at the beginning of this section) the object constructed is in some sense a completion of those constructed to obtain semilinear orders in 12.2 and 12.3.

The groups $\mathrm{Aut}(\Sigma, C)$ and $\mathrm{Aut}(P, \leq)$ are equal as abstract groups if Σ consists of *all* the maximal chains of P. But they are different as permutation groups as they act on different sets. In most cases,

Aut (Σ, C) is 2-transitive but not 3-transitive. This is true, for instance, when the C-relation is obtained from one of the countable 2-homogeneous semilinear orders listed by Droste (1985), or when it is built in a regular way from copies of **Z**. It is also not 2-primitive. To see this, fix some $\alpha \in \Sigma$ and declare two chains β and γ in $\Sigma \setminus \{\alpha\}$ to be equivalent if they meet α at the same node of P, that is, if $\alpha \cap \beta = \alpha \cap \gamma$. Then this relation is a proper $(\text{Aut}\,(\Sigma, C))_\alpha$-congruence on $\Sigma \setminus \{\alpha\}$.

Under certain conditions on C, the group $\text{Aut}\,(\Sigma, C)$ has many Jordan sets. Theorem 14.9 of Adeleke & Neumann (1996c) gives a complete list of Jordan sets in this category. Assume for example that $(P, \leq)$ is a countable 2-homogeneous semilinear order, and Σ is a set of maximal chains of P with union P. Then the set Γ of all chains in Σ passing through a fixed node of P is a Jordan set for $\text{Aut}\,(\Sigma, C)$. The set $\Sigma \setminus \Gamma$ is also a Jordan set. Define an equivalence relation E_a on the set of chains in Σ passing through a as follows: if $\alpha, \beta \in \Sigma$ and $a \in \alpha \cap \beta$, put $\alpha E_a \beta$ if there exists $y \in P$ such that $y > a$ and $y \in \alpha \cap \beta$. The E_a-classes in Σ are also called *cones* at a. The set of cones at a fixed node, say a, of P is a Jordan set. For $\alpha \in \Sigma$ let $I \subset P$ be an interval of α, and suppose that for $x, y \in P$, if $x \in I$ and $x > y$ then $y \in I$. Then the set $\{\gamma \in \Sigma \mid \alpha \cap \gamma \subseteq I\}$ is called a *lower section* of the C-relation. In the countable 2-homogeneous case for $(P, \leq)$, lower sections are Jordan sets. We shall have occasion to refer to these objects in the next chapter.

12.3   General betweenness relations

Let us start with a semilinear order $(P, \leq)$. We ignore the natural orientation and order imposed by the semilinear order and define a ternary relation B on P, intended to capture the notion of betweenness in an acyclic or tree-like object, by declaring that $B(x; y, z)$ if one necessarily has to pass through x to go from y to z or vice-versa. That is, x lies in the path between y and z (see Fig. 10). This idea is a generalisation of the notion of a linear betweenness relation that we have already defined in the last chapter (cf. Defn. 11.6). We formalise this idea in the definition below.

Definition 12.7 A ternary relation B defined on the set P is called a *B-relation* if

B1 $B(x; y, z) \Rightarrow B(x; z, y)$;

B2 $B(x; y, z) \land B(y; x, z) \Leftrightarrow x = y$;

B3 $B(x; y, z) \Rightarrow B(x; y, w) \lor B(x; z, w)$.

If, in addition, it satisfies

B4 $\neg B(x; y, z) \Rightarrow (\exists w \neq x)(B(w; x, y) \land B(w; x, z))$

then B is called a *general betweenness relation*.

Note that B1, B2 and B3 of the above definition are identical to b1, b2 and b3 of Definition 11.6. If, whenever $x \neq y$, there exists $z \neq x, y$ such that $B(z; x, y)$ then the relation is said to be *dense*.

A semilinear order can be recovered from a general betweenness relation in a very natural way. Given (P, B) and $x \in P$, we define a relation $\leq_x$ on the set $P \setminus \{x\}$ by the rule

$$y \leq_x z :\Leftrightarrow B(y; x, z).$$

It is easy to see that $\leq_x$ is a partial ordering on $P \setminus \{x\}$. To see that this also defines a semilinear order (albeit on a different set) we refer to Theorem 15.7 of Adeleke & Neumann (1996c).

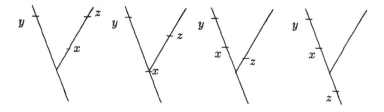

Figure 10: $B(x; y, z)$

Clearly, $\mathrm{Aut}\,(P,\leq) \leq \mathrm{Aut}\,(P,B)$. Therefore, $\mathrm{Aut}\,(P,B)$ is a Jordan group if $\mathrm{Aut}\,(P,\leq)$ is one. Suppose now that $(P,\leq)$ is a countable 2-homogeneous semilinear order, like $P(\mathbb{Q},2,+)$ which we have constructed earlier. Then $\mathrm{Aut}\,(P,\leq) < \mathrm{Aut}\,(P,B)$. This is because when we forget the ordering on P, the set $\{x \mid a \not\leq x\}$ looks exactly like a cone at a. There is a notion of a 'cone at a' for (P,B) such that (P,B) has one more cone at the node a than $(P,\leq)$. Suppose there are k cones at a and the automorphism group, $\mathrm{Aut}\,(D,\leq)$, of each cone D at a is H. Then the stabiliser of a in $\mathrm{Aut}\,(P,B)$ is isomorphic to $H\,\mathrm{Wr}\,S_{k+1}$. This group is transitive on $P\setminus\{a\}$ as H and S_{k+1} are transitive on their respective sets. Hence $\mathrm{Aut}\,(P,B)$ is 2-transitive. But $\mathrm{Aut}\,(P,\leq)$ is not 2-transitive. Thus they cannot be the same. $\mathrm{Aut}\,(P,B)$ however, is not 2-primitive because the cones at a node a are equivalence classes for the stabiliser. Hence, it is not 3-transitive. Cones at a node, unions of cones at a node as well as unions of chains of cones are Jordan sets of the automorphism group.

12.4 D-relations

In a general betweenness relation defined in the last section, the order of an underlying semilinear order is not important. We want to define a relation D that corresponds to a B-relation just as a C-relation corresponds to a semilinear order. Elements of a C-relation (that is, maximal chains in a semilinear order) can be thought of as directions in a semilinear order. Likewise, elements of a D-relation can be thought of as *directions* of the B-relation (see Section 16 of Adeleke & Neumann (1996c) for a precise definition). In a B-relation, there will be maximal sets of nodes on which B induces a *linear* betweenness relation, and any such set determines two directions. The D-relations of interest to us lie on a set of infinite domain, and arise from a non-linear infinite betweenness relation. Directions are also sometimes called *ends* or *points at infinity* and are then defined to be equivalence classes of rays or half lines.

The group $\mathrm{Aut}\,(\Omega, D)$ can be 3-transitive as, for certain examples, it is possible to map any set of three directions to any other set of three directions. Since any three directions meet at a point, such a mapping will map the meeting point of the first set to that of the second. It is not however, 4-transitive because any automorphism fixing the three directions would also fix the point at which they meet. Thus the relation D must be quaternary.

Let Ω be the set of directions of a general betweenness relation (P, B). Any two distinct directions determine a unique path via the underlying betweenness relation. We call such a path a *line*. We can then define a quaternary relation D on the set of directions Ω, by saying that for distinct $\alpha, \beta, \gamma, \delta$ the relation $D(\alpha, \beta; \gamma, \delta)$ holds if and only if the line through α and β does not meet the line through δ and γ (see Fig. 11).

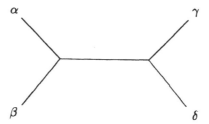

Figure 11: $D(\alpha, \beta; \gamma, \delta)$

It can be shown (cf. Adeleke & Neumann 1996c, Thm. 26.4) that given a D-relation D on a set Ω, there is a general betweenness relation (Λ, B) such that Ω is a set of directions of (Λ, B) and the relation D is derived from it as described in the last paragraph. It is possible however that Λ is merely 'dense', in a natural sense, in the set of directions. This is an analogue of Theorem 12.5.

Note also that for any $\alpha \in \Omega$ if we define a relation C on $\Omega \setminus \{\alpha\}$ by $C(\beta; \delta, \gamma)$ if and only if $D(\alpha, \beta; \gamma, \delta)$, then $(\Omega \setminus \{\alpha\}, C)$ is a C-relation. Conversely,,, given a C-relation defined on a set Ω, Theorem 23.4 of Adeleke & Neumann (1996c) shows how to explicitly define a D-relation on the set $\Omega \cup \{\infty\}$, with $\infty \notin \Omega$. This is the basis of the formal definition of a D-relation which we state below.

Definition 12.8 A quaternary relation $D(\alpha, \beta; \gamma, \delta)$ on Ω is a D-*relation* if

D1 $D(\alpha, \beta; \gamma, \delta) \Rightarrow D(\beta, \alpha; \gamma, \delta) \wedge D(\alpha, \beta; \delta, \gamma) \wedge D(\gamma, \delta; \alpha, \beta)$;

D2 $D(\alpha, \beta; \gamma, \delta) \Rightarrow \neg D(\alpha, \gamma; \beta, \delta)$;

D3 $D(\alpha, \beta; \gamma, \delta) \Rightarrow D(\epsilon, \beta; \gamma, \delta) \vee D(\alpha, \beta; \gamma, \epsilon)$;

D4 $(\alpha \neq \gamma \wedge \beta \neq \gamma) \Rightarrow D(\alpha, \beta; \gamma, \gamma)$;

D5 $(\alpha, \beta, \gamma \text{ distinct }) \Rightarrow (\exists \delta)(\gamma \neq \delta \wedge D(\alpha, \beta; \gamma, \delta))$.

As already mentioned $\mathrm{Aut}\,(\Omega, D)$ may be 3-transitive. But it is not 3-primitive, as the point stabiliser preserves a C-relation and is therefore not 2-primitive. Note that a two-point stabiliser of the group $\mathrm{Aut}\,(\Omega, D)$ fixes a line. Call two directions equivalent with respect to the line if they meet in the same point on that line. The equivalence classes are then the blocks of imprimitivity of the two-point stabiliser of $\mathrm{Aut}\,(\Omega, D)$. The group $\mathrm{Aut}\,(\Omega, D)$ is a Jordan group, because its one-point stabiliser is the automorphism group of a suitably homogeneous C-relation. A complete classification of proper Jordan sets for $\mathrm{Aut}\,(\Omega, D)$ is given in Theorem 28.6 of Adeleke & Neumann (1996c).

In the theory of infinite graphs too, there is a notion of an 'end'. Möller (1992) has shown that if Γ is an infinite graph with infinitely many ends whose automorphism group is transitive on the vertices and the ends, and all whose vertices have finitely many neighbours, then there is a natural D-relation on which $\mathrm{Aut}\,(\Gamma)$ acts.

12.5 Primitive groups with primitive Jordan sets

Let G be a permutation group acting primitively on a set Ω. Suppose G has a primitive Jordan set $\Sigma \subseteq \Omega$. We have already seen examples of such groups. The group induced on an open interval in $(\mathbb{Q}, \leq)$ is highly homogeneous and hence an open interval in $\mathbb{Q}$ is a primitive Jordan set of the primitive automorphism group $\mathrm{Aut}\,(\mathbb{Q}, \leq)$. In a 2-homogeneous semilinear order, like $P(\mathbb{Q}, 2, +)$, every cone is a primitive Jordan set. The following theorem characterises the groups that occur as groups of automorphisms of the relational structures discussed in this chapter and in Section 11.3.

Theorem 12.9 (Adeleke & Neumann 1996a) *Suppose that G is a primitive permutation group that has primitive proper Jordan sets. If G is not highly transitive then it preserves a relation which is either one of the four types of linear relation described in Section 11.3 or one of the four types of relation described in this chapter.* $\square$

<u>Exercises:</u>

12(i) Let G be a group with a chain $\{G_i \mid i \in \mathbb{Q}\}$ of normal subgroups with trivial intersection, and define the relation $C(x; y, z)$ to hold on G if some coset of some G_i contains y and z but not x. Show that C is a C-relation, and is preserved by left and right group multiplication.

12(ii) Describe all Jordan sets for the automorphism group of the C-relation derived from the semilinear order $P(\mathbb{Q}, 2, +)$.

12(iii) Show that $P(\mathbb{Q}, 2, +)$ and the corresponding derived betweenness relation have oligomorphic automorphism groups.

12(iv) Describe how one can define a relation B on a semilinearly ordered set $(P, \leq)$ (in terms of $\leq$) so that it satisfies B1 – B4 of Definition 12.7.

12(v) Show that in Definition 12.4, the set Ω must be infinite.

12(vi) Given a 2-homogeneous semilinearly ordered set $(P, \leq)$,
    (a) show how one can define a D-relation on the set of maximal chains;
    (b) show how one can define a D-relation on the set of maximal chains with an extra point (corresponding to the downward direction) adjoined.

12(vii) Show that if $a \in P(\mathbb{Q}, n, +)$ and $G := \text{Aut}(P(\mathbb{Q}, n, +))$, then $G_a \cong (G \,\text{Wr}\, S_n) \times (G \,\text{Wr}\, S_{n-1}) \,\text{Wr}\, \text{Aut}(\mathbb{Q}, <)$, where G_a denotes the stabiliser of a in G.

Chapter 13

Classification Theorems

We have already seen some classification theorems in the earlier chapters. Theorem 10.16 classifies primitive groups with cofinite Jordan sets. Theorem 11.12 classifies all infinite permutation groups which are highly homogeneous but not highly transitive. Theorem 12.9 classifies infinite primitive permutation groups which have proper primitive Jordan sets. We have also seen many different examples of Jordan groups in the last three chapters. In this chapter we tie together all the different classes of examples by a theorem (due to Adeleke & Macpherson) which extends Theorems 11.12 and 12.9 and classifies all infinite primitive Jordan groups G. Theorem 11.12 tells us that if G is highly homogeneous but not highly transitive then it preserves a relation of one of the four types described in Section 11.3. If G has primitive Jordan sets and is not highly homogeneous then by Theorem 12.9 it follows that G preserves a relation of one of the four types described in Chapter 12. The classification theorem of Adeleke & Macpherson tells us that if G is not one of the types mentioned above, and is not highly transitive, then it must either be an automorphism group of a Steiner k-system (cf. Sec. 11.2) or it must preserve limit structures of certain specified kinds.

13.1 Statement of the theorems

The first step towards the classification of all infinite primitive Jordan groups is Theorem 11.12 due to Cameron which classifies all highly homogeneous but not highly transitive permutation groups. We state

131

the theorem again in a slightly different way.

Theorem 13.1 (Cameron 1976) *Let G be a permutation group on an infinite set Ω, and suppose that G is highly homogeneous but not highly transitive. Then G preserves (and has the same orbits on finite ordered sets as the full automorphism group of) one of the following relational structures:*

(i) *a dense linear order;*

(ii) *a dense linear betweenness relation;*

(iii) *a dense cyclic order;*

(iv) *a dense cyclic separation relation.*

In particular, the above theorem implies that G is at most 3-transitive.

The next step towards the classification theorem of primitive Jordan groups is Theorem 12.9 which classifies all primitive Jordan groups with primitive Jordan sets. We state the theorem again in this setting.

Theorem 13.2 (Adeleke & Neumann 1996a) *Suppose that G is a primitive permutation group that has primitive proper Jordan sets. If G is not highly transitive then there is a G-invariant relation R on Ω which is one of* (i) – (iv) *of Theorem 13.1 or is one of the following:*

(v) *a dense semilinear order;*

(vi) *a dense general betweenness relation;*

(vii) *a C-relation;*

(viii) *a D-relation.*

The assumption that G has a proper primitive Jordan set imposes very strong conditions on the group. Adeleke and Macpherson, in their recent work, have classified those primitive Jordan groups which do not have primitive Jordan sets and have thereby completed the classification of all infinite primitive Jordan groups. We use the word 'classification' in a broad sense – Jordan groups are put into classes, but are not determined up to isomorphism type by known invariants.

Theorem 13.3 (Adeleke & Macpherson 1995) **The Classification Theorem:** *Suppose that G is an infinite primitive Jordan group which is not highly transitive. Then, either G preserves one of the relations* (i) – (viii) *mentioned in Theorem 13.2 above, or one of the following holds:*

(ix) *G is a group of automorphisms of a Steiner k-system for some $k \geq 2$;*

(x) *none of* (i) – (ix) *holds, and G preserves a limit of a sequence of Steiner systems, general betweenness relations, or D-relations.*

Remarks:

I In (i) – (viii) of the above list, the groups are always at most 3-transitive.

II If (G, Ω) is of type (ix) or (x) then it has no proper primitive Jordan set. Also if it is of any of the types (i) – (vi) but not of types (vii), (viii) or (ix) then it has a proper primitive Jordan set.

III A slightly stronger statement of (vii) is true. We can insist that either all cones are Jordan sets, or all lower sections, defined in terms of the underlying semilinear order (cf. Sec. 12.2), are Jordan sets.

IV For any k, the examples in (ix) can be chosen such that G is k-transitive but not $(k + 1)$-transitive. We shall construct examples of such groups in Chapter 15.

V The statement of (x) is vague. We shall not try to be more explicit, or even define what we mean by these limit structures as they are very complicated objects and we shall not discuss them any more either here or in later chapters. A full account can be found in Adeleke & Macpherson (1995) and also in Macpherson (1994). The examples of groups in this category are pathological and there are many open questions. However, Adeleke (1995, preprint) has shown that these limit objects do exist.

Since the proof of Theorem 13.3 is long and highly technical in parts, no useful purpose will be served in reproducing it here. The interested reader is referred to Adeleke & Neumann (1996a) for a proof of Theorem 13.2, and to Adeleke & Macpherson (1995) or to a survey by Macpherson (1994) for detailed proofs of the complete classification. In this chapter we propose to state a more detailed version of the classification theorem, to explain (in Sections 13.2 and 13.3) some of the key concepts and ideas that have been used, and then to give, in the last section, a brief overview of the different steps involved in the proof. As there is much to be filled in by the reader in the following three sections, we do not have any formally stated exercises in this chapter.

The following version of the classification theorem indicates how different degrees of transitivity correspond to different invariant structures and also states where in Adeleke & Macpherson (1995) one can find a proof of each separate claim. The types (i) – (x) below all refer to the list in the last three theorems. For example, we say that (G, Ω) is of type (i) if it preserves a dense linear order.

Theorem 13.4 *Let (G, Ω) be an infinite primitive Jordan group which is not highly transitive. Then one of the following holds:*

- *(1) (G, Ω) is highly homogeneous but not 2-transitive, and is of type (i) (Theorem 3.1.1);*

- *(2) (G, Ω) is not 2-homogeneous, and is of type (v) or (vii) (Theorem 3.1.1);*

- *(3) (G, Ω) is 2-transitive but not 2-primitive, and is of type (ii), (vi), (vii), (viii) or (ix) (Theorems 3.1.2, 5.6.1, 5.1.2);*

- *(4) (G, Ω) is 2-primitive but not 3-transitive, and is of type (iii), (viii) or (x) (a limit of general betweenness relations or D-relations) (Theorems 3.1.3, 5.7.2);*

- *(5) (G, Ω) is 3-transitive but not 3-primitive, and is of type (iv), (viii), (ix) or (x) (a limit of Steiner systems) (Theorems 3.1.4, 5.8.2);*

- *(6) for some $k \geq 4$, (G, Ω) is k-transitive but not k-primitive, and is of type (ix) or (x) (a limit of Steiner systems) (Theorems 5.8.3, 5.8.4).*

13.2 Some recognition theorems

In this section, we discuss some basic theorems which help in the recognition (and also elimination) of the relations in the list. Almost all the material in this section has been taken from Section 34 of Adeleke & Neumann (1996c). Given a subset Σ of Ω and a group G acting on Ω, we call the sets Σ^g for $g \in G$ the *translates* of Σ. Recall that by Lemma 10.3 if Σ is a Jordan set, then so is the translate Σ^g.

Theorem 13.5 *Let G be primitive on Ω and suppose there is $\Sigma \subseteq \Omega$ with $\Sigma \neq \emptyset, \Omega$ such that for all $g \in G$*

$$\Sigma \subseteq \Sigma^g \ or \ \Sigma^g \subseteq \Sigma.$$

Then

(i) *there is a G-invariant linear order $\leq$ on Ω such that Σ is an initial segment of $(\Omega, \leq)$.*

(ii) *if Σ is a Jordan set, then G is highly homogeneous on Ω.*

Proof: (i) Define $\alpha \leq \beta$ if for all $g \in G$, $\beta \in \Sigma^g \Rightarrow \alpha \in \Sigma^g$. Clearly, $\leq$ is reflexive, transitive and G-invariant. To prove anti-symmetry, define a relation $\sim$ on Ω by declaring $\alpha \sim \beta$ if and only if $\alpha \leq \beta \wedge \beta \leq \alpha$. Equivalently,

$$\alpha \sim \beta \ \text{if and only if} \ (\forall g \in G)(\alpha \in \Sigma^g \Leftrightarrow \beta \in \Sigma^g).$$

This relation is easily seen to be an equivalence relation. It is also G-invariant. But G is primitive. So $\sim$ must be either trivial or universal. Since Σ is a proper subset of Ω we can choose $\alpha \in \Sigma$, $\beta \notin \Sigma$ and g to be the identity in G. Then the condition $\alpha \in \Sigma^g \Leftrightarrow \beta \in \Sigma^g$ is violated. So $\sim$ is not universal, and the equivalence classes are singletons as required.

To prove that the order is linear, we have to show that if $\alpha \neq \beta$ then either $\alpha < \beta$ or $\beta < \alpha$. Since the $\sim$ classes are singletons, we can assume, without loss of generality, that there exists $g \in G$ such that $\alpha \in \Sigma^g$ but $\beta \notin \Sigma^g$. In this case, we claim that $\alpha < \beta$. Suppose not. Then $\alpha \not\leq \beta$ and so, by the definition of $\leq$, there must exist $h \in G$ such that $\beta \in \Sigma^h$ but $\alpha \notin \Sigma^h$. Then $\Sigma^g \setminus \Sigma^h$ and $\Sigma^h \setminus \Sigma^g$ are non-empty, which is a contradiction to our initial assumption on Σ.

It only remains to show that Σ is an initial segment of Ω. But this is easy because if not, we can choose $\alpha < \beta$ with $\alpha \notin \Sigma$ and $\beta \in \Sigma$, contrary to the definition of $\leq$.

(ii) We shall show that, for all $k \in \mathbf{N}$, the group G is order k-transitive (cf. p. 117) on Ω. This is equivalent to showing that G is transitive on the set

$$\{(\alpha_1, \alpha_2, \ldots, \alpha_k) \in \Omega^k \mid \alpha_1 < \alpha_2 < \ldots < \alpha_k\}.$$

The proof is by induction on k.

For $k = 1$ the proof follows from the transitivity of G. So let us assume that G is order k-transitive for $k \geq 1$ and consider two chains

$$\alpha_0 < \alpha_1 < \ldots < \alpha_k \text{ and } \beta_0 < \beta_1 < \ldots < \beta_k.$$

Then by inductive hypothesis, there is an element $g \in G$ such that $\beta_i^g = \alpha_i$ for $1 \leq i \leq k$. Since g is order-preserving, $\beta_0^g < \alpha_1$. Also $\alpha_0 < \alpha_1$. Let $\beta := \max\{\beta_0^g, \alpha_0\}$. By definition of the relation $\leq$ there is some translate Σ^* of Σ which contains β and hence also β_0^g and α_0 but not α_1. Since Σ^* is also a Jordan set, there exists $h \in G_{(\Omega \setminus \Sigma^*)}$ that maps β_0^g to α_0. Then $\beta_i^{gh} = \alpha_i$ for $0 \leq i \leq k$. $\square$

For the rest of this section, we shall not give complete proofs of the theorems that follow but only hints for the proof. This is because, given the hints, the proofs are, in most cases, just routine verification of the axioms. Also because the proofs can be found, complete in every detail, in §34 of Adeleke & Neumann (1996c). Recall that two subsets Γ, Δ of a set Σ are said to form a *typical pair of sets* or simply a *typical pair* if all the sets $\Gamma \setminus \Delta$, $\Delta \setminus \Gamma$ and $\Gamma \cap \Delta$ are non-empty (cf. Ex. 10(viii)). In this case, we sometimes also say that the pair (Γ, Δ) is *typical*.

Theorem 13.6 *Assume that G is primitive on Ω and that there is $\Sigma \subseteq \Omega$ with $\Sigma \neq \emptyset, \Omega$ such that*

(i) *for every $g \in G$, the pair (Σ, Σ^g) is not typical; and*

(ii) *there exist distinct $\alpha_0, \beta_0 \in \Omega$ such that*
$$(\forall g \in G)(\beta_0 \in \Sigma^g \Rightarrow \alpha_0 \in \Sigma^g).$$

Then there is a G-invariant linear or (lower) semilinear order on Ω.

Hint: Define $\beta \le \alpha$ if for all $g \in G$, $\beta \in \Sigma^g \Rightarrow \alpha \in \Sigma^g$. $\square$

Theorem 13.7 *Assume G is transitive on Ω and there is $\Sigma \subseteq \Omega$ such that*

(i) $|\Sigma| > 1$;

(ii) *for every $g \in G$, the pair (Σ, Σ^g) is not typical;*

(iii) $(\forall \alpha, \beta \in \Omega)(\exists g \in G)(\alpha, \beta \in \Sigma^g)$; *and*

(iv) $(\forall \alpha, \beta \in \Omega)(\alpha \neq \beta \Rightarrow (\exists g \in G)(\alpha \in \Sigma^g \wedge \beta \notin \Sigma^g))$.

Then there is a G-invariant C-relation on Ω.

Hint: Define a ternary relation C on Ω by $C(\alpha; \beta, \gamma)$ if there exists $g \in G$ such that $\beta, \gamma \in \Sigma^g$ but $\alpha \notin \Sigma^g$. Show that this is a C-relation. $\square$

Theorem 13.8 *Assume G is 2-transitive on Ω and there is $\Sigma \subseteq \Omega$ such that*

(i) *for any $g \in G$, if (Σ, Σ^g) is a typical pair then $\Sigma \cup \Sigma^g = \Omega$;*

(ii) *there are $x, y \in G$ such that the pairs (Σ, Σ^x) and $(\Omega \backslash \Sigma, (\Omega \backslash \Sigma)^y)$ are both typical;*

(iii) *there exist distinct $\alpha_0, \beta_0, \gamma_0 \in \Omega$ such that*
$(\forall g \in G)(\beta_0, \gamma_0 \in \Sigma^g \Rightarrow \alpha_0 \in \Sigma^g)$.

Then there is a G-invariant dense betweenness relation on Ω.

Hint: Define a ternary relation B on Ω by declaring that $B(\alpha; \beta, \gamma)$ if and only if every translate of Σ which contains β and γ also contains α. Then show that this is a B-relation. $\square$

Theorem 13.9 *Assume G is transitive on Ω and there is $\Sigma \subseteq \Omega$ such that*

(i) $|\Sigma| > 1, |\Omega \backslash \Sigma| > 2$;

(ii) *for all $g \in G$, if (Σ, Σ^g) is a typical pair then $\Sigma \cup \Sigma^g = \Omega$;*

(iii) $(\forall \alpha, \beta, \gamma \in \Omega)((\alpha, \beta, \gamma \text{ distinct}) \Rightarrow$
$(\exists g \in G)(\beta, \gamma \in \Sigma^g \wedge \alpha \notin \Sigma^g))$.

Then there is a G-invariant D-relation on Ω.

Hint: Define the relation D on Ω by declaring that $D(\alpha, \beta; \gamma, \delta)$ holds if there exists $g \in G$ such that $(\alpha, \beta \in \Sigma^g \wedge \delta, \gamma \notin \Sigma^g)$ or $(\alpha, \beta \notin \Sigma^g \wedge \delta, \gamma \in \Sigma^g)$. □

Theorem 13.10 *Suppose G is 2-transitive but not 2-primitive on Ω. Let $\alpha \in \Omega$ and let Δ be a block of imprimitivity of G_α on $\Omega \setminus \{\alpha\}$. Let $\beta \in \Delta$. Suppose there is $g \in G$ such that $\alpha^g = \beta$ and $\beta^g = \alpha$ and such that $\Delta \cup \{\alpha\}^g = \Delta \cup \{\alpha\}$. Then there is a G-invariant Steiner 2-system on Ω.*

Hint: The Steiner blocks are the G-translates of $\Delta \cup \{\alpha\}$. □

We end this section by stating a theorem for recognising Steiner k-systems. The proof is left as an exercise.

Theorem 13.11 *Let n be a natural number greater than two and let Γ be an infinite set containing an element α. Suppose that (H, Γ) is an n-transitive but not $(n+1)$-transitive permutation group such that H_α preserves a Steiner $(n-1)$-system on $\Gamma \setminus \{\alpha\}$ with a Steiner block Δ. Suppose also that there is an $x \in H$ such that for some $\beta \in \Delta$ we have $\alpha^x = \beta$, $\beta^x = \alpha$ and $(\Delta \cup \{\alpha\})^x = (\Delta \cup \{\alpha\})$. Then H preserves a Steiner n-system on Γ whose blocks are the H-translates of $\Delta \cup \{\alpha\}$.* □

13.3   Some basic lemmas

This section is devoted to a few basic lemmas which are used repeatedly, and to great effect, in the proof of the classification.

The proof of the classification theorem is by induction on the degree of transitivity of the Jordan group. The following lemma is crucial in getting the induction started in the case where there is a primitive Jordan set.

Lemma 13.12 (Adeleke & Neumann 1996a, Lemma 4.5) *If (Σ_1, Σ_2) is a typical pair of proper Jordan sets such that Σ_1 is primitive and Σ_2 is k-homogeneous, where $k \geq 1$ and $2k \leq |\Sigma_2|$, then their union $\Sigma := \Sigma_1 \cup \Sigma_2$ is $(k+1)$-homogeneous.*

Proof: Let $H_1 := G_{(\Omega \setminus \Sigma_1)}, H_2 := G_{(\Omega \setminus \Sigma_2)}$ and $H := G_{(\Omega \setminus \Sigma)}$. The family $\{\Sigma_1^x \mid x \in H_2\}$ is connected since all its members contain $\Sigma_1 \setminus \Sigma_2$, and its union is Σ. By Corollary 10.10, therefore Σ is a primitive Jordan set. Similarly, we see that Σ is k-homogeneous.

If $\Sigma_2 \setminus \Sigma_1$ is finite, then Σ_1 is a cofinite primitive Jordan set for H in Σ and so is improper (by Theorem 10.14). In this case therefore, H is at least 2-primitive on Σ, and so if $\alpha \in \Sigma_1 \setminus \Sigma_2$ then H_α is primitive on $\Sigma \setminus \{\alpha\}$, in which Σ_2 is a k-homogeneous Jordan set. By Theorem 10.11, the stabiliser H_α is k-homogeneous, and so by Theorem 3.17 the group H is $(k+1)$-homogeneous.

We may suppose therefore that $\Sigma_2 \setminus \Sigma_1$ is infinite (although all we really need is that $|\Sigma_2 \setminus \Sigma_1| \geq k$). Let Δ_0 be a k-subset of Σ_2 and δ_0 an element of $\Sigma_1 \setminus \Sigma_2$. We need to show that any $(k+1)$-subset Δ of Σ may be mapped to $\Delta_0 \cup \{\delta_0\}$ by some element of H. Let $\delta_1 \in \Delta$ and let $\Delta_1 := \Delta \setminus \{\delta_1\}$. Since H is k-homogeneous, we may certainly map Δ_1 to a k-set consisting of $k-1$ points of $\Sigma_2 \setminus \Sigma_1$ and one point of Σ_1. So we may suppose that Δ either consists of one point of Σ_1 and k points of $\Sigma_2 \setminus \Sigma_1$ or consists of two points α and β of Σ_1 and $k-1$ points of $\Sigma_2 \setminus \Sigma_1$. But in the latter case, there exists $h_1 \in H_1$ such that one of $\alpha^{h_1}, \beta^{h_1}$ lies in $\Sigma_1 \setminus \Sigma_2$ while the other lies in $\Sigma_1 \cap \Sigma_2$ by Lemma 4.7. Then Δ^{h_1} consists of one point of $\Sigma_1 \setminus \Sigma_2$ and k points of Σ_2, and since Σ_2 is k-homogeneous, there exists $h_2 \in H_2$ such that $\Delta^{h_1 h_2}$ consists of the one point from $\Sigma_1 \setminus \Sigma_2$ and k points of $\Sigma_2 \setminus \Sigma_1$. Thus in any event, we may suppose that Δ contains one point δ from Σ_1 and k points from $\Sigma_2 \setminus \Sigma_1$. There exists $g_1 \in H_1$ mapping δ to δ_0, and there exists $g_2 \in H_2$ mapping $\Delta \setminus \{\delta\}$ to Δ_0, so that $g_1 g_2$ is an element of H which maps Δ to $\Delta_0 \cup \{\delta_0\}$ as required. $\square$

From the last lemma and Theorem 10.11 the following corollary is immediate.

Corollary 13.13 *Suppose G is primitive and has a typical pair of Jordan sets (Σ_1, Σ_2) in Ω. Then*

(i) *if one of Σ_1 and Σ_2 is primitive, then G is 2-homogeneous on Ω;*

(ii) *if Σ_1 and Σ_2 are both k-homogeneous, where $2k \leq \min(|\Sigma_1|, |\Sigma_2|)$ and $k \geq 2$, then G is $(k+1)$-homogeneous on Ω.* $\square$

We have already shown in Corollary 10.6 that given distinct points $\omega, \alpha_1, \alpha_2, \ldots, \alpha_n \in \Omega$ if there is a Jordan set in (G, Ω) containing ω

and omitting $\alpha_1, \alpha_2, \ldots, \alpha_n$, then there is a unique maximal such Jordan set. We denote this set by $MJ(\alpha_1, \alpha_2, \ldots, \alpha_n/\omega)$, ($MJ$ for maximal Jordan). We need to use such sets in the proof of the following lemma.

Lemma 13.14 (Adeleke & Macpherson 1995, Lemma 2.2.8)
Let (G, Ω) be an infinite Jordan group and let (Γ_1, Γ_2) be a typical pair of Jordan sets with union Γ. Then there is a unique maximal $G_{(\Omega \backslash \Gamma)}$-congruence ρ on Γ. Furthermore, either each of $\Gamma_1 \backslash \Gamma_2, \Gamma_2 \backslash \Gamma_1$ is a ρ-block, in which case $G_{(\Omega \backslash \Gamma)}$ is 2-transitive on Γ/ρ, or $G_{(\Omega \backslash \Gamma)}$ preserves a linear order, or a linear or general betweenness relation on Γ/ρ.

Hint: We can describe ρ partly as follows. Let $\alpha_1 \in \Gamma_1 \backslash \Gamma_2$ and $\alpha_2 \in \Gamma_2 \backslash \Gamma_1$. A typical ρ-class is the set of $\delta \in \Gamma \backslash \Gamma_2$ such that $MJ(\alpha_1/\alpha_2)$ is the same as $MJ(\delta/\alpha_2)$. □

13.4 Outline of the proof of Theorem 13.3

We begin with a primitive Jordan group G acting (faithfully) on an infinite set Ω. Let Σ be a proper Jordan set of G. We may suppose that Σ is infinite, since otherwise there is a finitary permutation and G is highly transitive (cf. Thm. 6.8 and Cor. 6.5). Also if Σ satisfies

$$(\forall g \in G)(\Sigma \subseteq \Sigma^g \text{ or } \Sigma^g \subseteq \Sigma)$$

then G is highly homogeneous and there is a G-invariant linear order on Ω by Theorem 13.5.

So suppose the above condition does not hold. If Σ is such that for every $g \in G$, the pair (Σ, Σ^g) is not typical, then we can apply Theorems 13.6 and 13.7 to get a semilinear order or a C-relation. Suppose (ii) of Theorem 13.6 holds. We then get a semilinear order by Theorem 13.6. If not, then (iv) of Theorem 13.7 holds. Part (iii) of that theorem is a consequence of Lemma 10.7. Thus, we get a C-relation in this case.

In fact, one can show (cf. Adeleke & Macpherson 1995, Thm. 3.1.1) that if G is primitive but not 2-transitive, then it preserves either a linear order, a semilinear order or a C-relation.

So we may assume that G is at least 2-transitive. We will also assume that there is $g \in G$ such that the pair (Σ, Σ^g) is typical. Then $\Sigma \cup \Sigma^g$ is also a Jordan set by Lemma 10.5. If Σ is a primitive Jordan set, by Lemma 13.12 it follows that $\Sigma \cup \Sigma^g$ will be a Jordan set (not necessarily proper) of a higher degree of homogeneity. Arguments of this sort are used in Adeleke & Neumann (1996a) to prove Theorem 13.2. So we can also assume that G has no primitive Jordan sets.

When G is 2-transitive and not 3-transitive, we need to consider two cases:

Case (i): There exist distinct $\alpha, \beta, \gamma \in \Omega$ such that no Jordan set contains α and omits β and γ. In this case, we apply Theorem 5.1.2 of Adeleke & Macpherson (1995) to obtain G-invariant Steiner systems (cf. Thm. 13.10) or C-relations.

Case(ii): Otherwise, for any distinct $\alpha, \beta, \gamma \in \Omega$ there is a unique Jordan set, denoted by $MJ(\alpha, \beta/\gamma)$, which is maximal subject to containing γ and excluding α and β. Since G is not 3-transitive, $MJ(\alpha, \beta/\gamma)$ is a proper Jordan set. As mentioned before, we are assuming that there is a typical pair of Jordan sets. Furthermore, if the union of every typical pair of proper Jordan sets is the whole of Ω then we get a G-invariant C- or D-relation on Ω (cf. Thms. 13.7 & 13.9). So we can also suppose that this does not hold.

Let Γ_1, Γ_2 be a typical pair of Jordan sets with union $\Gamma \neq \Omega$. Pick $\beta \in \Gamma_1 \setminus \Gamma_2$, $\gamma \in \Gamma_2 \setminus \Gamma_1$ and $\alpha \in \Omega \setminus \Gamma$. Then the pair $(MJ(\alpha, \beta/\gamma), MJ(\alpha, \gamma/\beta))$ is typical. Call their union Θ_0. Then this is a Jordan set. Given a point $\alpha_1 \notin \Theta_0$, we can similarly obtain $MJ(\alpha_1, \gamma/\beta)$ and $MJ(\alpha_1, \beta/\gamma)$. By a careful choice of α_1, these Jordan sets can be made strictly bigger than $MJ(\alpha, \gamma/\beta)$ and $MJ(\alpha, \beta/\gamma)$ respectively. Call their union Θ_1. Continue this process for points $\alpha_2, \alpha_3, \ldots, \alpha_j, \ldots$ such that $\alpha_j \notin \Theta_{j-1}$. We obtain a chain of bigger and bigger Jordan sets Θ_j and find a maximal chain of such Jordans sets. Then the union of all the members of this chain must be the whole of Ω. We can then apply Lemma 13.14 to obtain either a limit of D-relations or general betweenness relations preserved by G.

To apply induction, we show that if a Jordan group G is

(a) 3-primitive then it is 4-transitive;

(b) 3-transitive and not 3-primitive then it is one of the known types.

We define k to be the maximal degree of transitivity of G. The proof of the theorem is inductive after the first few steps (induction starts at $k \geq 3$), for if (G, Ω) is k-transitive, then the stabiliser in G of a point is $(k-1)$-transitive on the remaining points. Therefore we only have to examine transitive extensions of already known permutation groups. (A transitive permutation group (G, Ω) is said to be a *transitive extension* of (H, Σ) if there is $\alpha \in \Omega$ such that $\Sigma = \Omega \setminus \{\alpha\}$ and the permutation groups $(G_\alpha, \Omega \setminus \{\alpha\})$ and (H, Σ) are isomorphic.) And we can show that for k sufficiently large, G preserves either a Steiner k-system or a limit of Steiner $(k-1)$-systems. This finishes the proof.

Chapter 14

Homogeneous Structures

In this chapter we take a break from the study of Jordan groups and look at model theory – a theory arising from set theory and the foundations of mathematics, and which has many connections with the study of permutation groups. An excellent reference for model theory is Hodges (1993), and our notation and terminology is based on it. An abridged version has also recently appeared (Hodges, 1997). Note that in the next two chapters, mappings will be written on the left of their arguments, contrary to the notation used in the earlier chapters.

14.1 Some model theory

All structures and languages considered here will be of first order, that is, variables will range over *elements* of a structure, rather than *subsets*. A *structure* or a *model* $\mathcal{M}$ is a set M (the *domain*), together with certain specified functions $M^n \rightarrow M$ and relations (subsets of M^n) for various $n \in \mathbf{N}$, and some distinguished elements called constants. The numbers n which occur are called the *arities* of the corresponding relations and functions. A relation (or function) of arity n will be called *n-ary*. The *cardinality* of a structure is the cardinality of its domain.

Each model is associated with a formal language, and we say $\mathcal{M}$ is an *$\mathcal{L}$-structure* if it is associated with the language $\mathcal{L}$. All our languages will have symbols for variables, propositional connectives (such as ∨ (for 'or'), ∧ (for 'and'), → (for 'implies'), ¬ (for negation)), together with the quantifiers ∀, ∃, and a binary relation symbol = which is always interpreted by equality. If $\mathcal{M}$ is an $\mathcal{L}$-structure, then $\mathcal{L}$ will have in addition relation and function symbols of the appropriate arities, and constant symbols, corresponding to the relations, functions and constants of $\mathcal{M}$. The language $\mathcal{L}$ is usually chosen with a class of structures (e.g. graphs, groups, ordered fields) in mind. We remark that constants can be regarded as functions of arity 0. A good example for languages and structures is the reals. The ordered field $(\mathbb{R}, <, +, \times, -, 0, 1)$ requires a language with a binary relation symbol, two binary and one unary function symbols, and two constant symbols. A language is called *relational* if it has no function or constant symbols. A *relational structure* is a structure over a relational language. For notational convenience, we will not distinguish between relation, function and constant symbols, and the corresponding relations, functions, and constants in a structure.

A *formula* is essentially a well-formed sequence of symbols in the language. There is a natural notion of the *scope* of a quantifier (∀ or ∃) in a formula. Given an occurrence of a variable, x say, in a formula, we say that the occurrence of x is *free* if it is not in the scope of any quantifier, and *bound* otherwise. A *sentence* is a formula in which every occurrence of every variable is bound. In general, we can only say that a formula is true (or false) in $\mathcal{M}$ *for a given interpretation* of its free variables by elements of $\mathcal{M}$, but as a sentence has no free variables, it is either true in $\mathcal{M}$ or false in $\mathcal{M}$. Model theory deals with the interaction between structures and languages: the extent to which the set of sentences true of a structure determine it; the extent to which properties of a structure are expressible in the language; and the complexity of the structure, in terms of the language.

A *substructure* or *submodel* $\mathcal{N}$ of $\mathcal{M}$ is determined by a set $N \subseteq M$ closed under all function symbols and constant symbols in $\mathcal{L}$. The relations and functions on $\mathcal{N}$ are the restrictions of those from $\mathcal{M}$. We write $\mathcal{N} \le \mathcal{M}$ to mean $\mathcal{N}$ is a substructure of $\mathcal{M}$. In the other direction, an *extension* of $\mathcal{M}$ is a model $\mathcal{N}$ in which $\mathcal{M}$ is a substructure.

We can also define the notion of isomorphism of structures. Two

$\mathcal{L}$-structures $\mathcal{M}$ and $\mathcal{N}$ are said to be *isomorphic* if there is a bijection between them which preserves relations, functions and constants in the usual sense. An *automorphism* of $\mathcal{M}$ is an isomorphism of $\mathcal{M}$ to itself. The set of all automorphisms of $\mathcal{M}$ forms a group under composition, denoted by $\mathrm{Aut}\,(\mathcal{M})$.

With the above terminology, we can make the following definition.

Definition 14.1 A structure $\mathcal{M}$ is $\aleph_0$-*categorical* (sometimes also called ω-*categorical*) if

(i) M is countably infinite, and

(ii) whenever $|M| = |N|$ and both $\mathcal{M}$ and $\mathcal{N}$ satisfy the same first order sentences, then they are isomorphic.

We have seen the axioms needed to define a dense linear order without endpoints in Section 9.1. All of them are naturally expressed as first-order sentences, in terms of the order relation $<$ of arity 2. In view of our discussion of the rationals in Chapter 9 and Cantor's Theorem, it follows that $(\mathbb{Q}, <)$ is $\aleph_0$-categorical. In the same chapter, we had also shown that $\mathrm{Aut}\,(\mathbb{Q}, <)$ is oligomorphic (Corollary 9.8). The following powerful result tells that these two properties are in fact equivalent in countably infinite structures. It also serves as a basic link between permutation group theory and model theory.

Theorem 14.2 (Ryll-Nardzewski 1959, Engeler 1959, Svenonius 1959) *Let $\mathcal{M}$ be a countably infinite structure. Then the following are equivalent.*

(i) $\mathcal{M}$ *is* $\aleph_0$-*categorical.*

(ii) $\mathrm{Aut}\,(\mathcal{M})$ *is oligomorphic.* $\square$

From now on, we shall not be too particular about distinguishing between a structure $\mathcal{M}$ and its domain M.

Definition 14.3 A relational structure M is *homogeneous* if

(i) M is countable, and

(ii) whenever U and V are finite substructures of M and $f : U \to V$ is an isomorphism, there is $\hat{f} \in \mathrm{Aut}\,(M)$ extending f.

Exercises:

14(i) We have seen in Exercise 12(iii) that the automorphism group of $P(\mathbb{Q}, 2, +)$ is oligomorphic. We invite the reader to write down the first order axioms which determine the countable models up to isomorphism (thereby witnessing the last theorem). Also to show, using a 'back and forth' argument, that any two countable models are isomorphic.

14(ii) Suppose that $\mathcal{L}$ is a finite relational language and M is a homogeneous $\mathcal{L}$-structure. Show that $\mathrm{Aut}\,(M)$ is oligomorphic.

14(iii) Show that $P(\mathbb{Q}, n, +)$ is homogeneous in the language with a C-relation, but that semilinear orders are not homogeneous. Also show that semilinear orders can be made homogeneous by adding finitely many extra relation symbols. Find them.

We remark here that given any permutation group G of countably infinite degree, G is dense in the automorphism group of a homogeneous structure – we just need to introduce a relation symbol for each orbit on ordered k-tuples, for all k.

We end this section with a word of caution. The word 'homogeneous' is used to mean something quite different in permutation group theory (cf. Defn. 3.16). This double usage of the word, though somewhat unfortunate, has become so standard that it is unavoidable. Even within model theory, the word 'homogeneous' has come to assume a variety of meanings. We will insist that the language we are working with is countable (which includes the finite case, in standard terminology). However, note that many of our theorems, like Theorem 14.4, have uncountable generalisations.

Examples :

14(a) $\mathrm{Aut}\,(\mathbb{Q}, <)$ is homogeneous as it is highly homogeneous in the sense of permutation groups.

14(b) A countably infinite set Ω with no relations at all is homogeneous because then $\mathrm{Aut}\,(\Omega) \cong \mathrm{Sym}\,(\mathbf{N})$.

14(c) A structure on a countably infinite set consisting of an equivalence relation with all classes of the same size can be easily shown to be homogeneous. The automorphism group of this structure is a wreath product of symmetric groups.

14.2 Building homogeneous structures

We now describe a very flexible technique for constructing homogeneous structures. The idea is to abstract the way in which $(\mathbb{Q}, <)$ is built from finite structures. It is due to Fraïssé. It provides a fertile source of interesting infinite permutation groups.

Let $\mathcal{L}$ be a relational language. Suppose that $\mathcal{C}$ is a non-empty class of finite $\mathcal{L}$-structures with the following properties:

(1) $\mathcal{C}$ is closed under isomorphism;

(2) $\mathcal{C}$ is closed under substructures *(Hereditary Property)*;

(3) whenever $A, B \in \mathcal{C}$ there is $D \in \mathcal{C}$ such that $A \leq D$ and $B \leq D$ *(Joint Embedding Property* or JEP);

(4) whenever $A, B_1, B_2 \in \mathcal{C}$ and $f_i : A \to B_i, i = 1, 2$ are embeddings, there exist $D \in \mathcal{C}$ and embeddings $g_i : B_i \to D, i = 1, 2$ such that for all $a \in A$ we have $g_1 \circ f_1(a) = g_2 \circ f_2(a)$ *(Amalgamation Property)*. See Fig. 12.

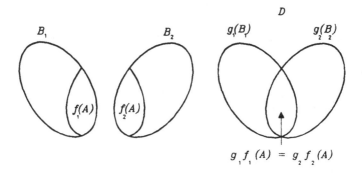

Figure 12: Amalgamation

Notes:

  I If D and the embeddings g_i defined in (4) above can be so chosen that $g_1(B_1) \cap g_2(B_2) = g_1 \circ f_1(A) = g_2 \circ f_2(A)$, then we say that C has the *Strong Amalgamation Property*.

  II The empty set is not considered as a structure in model theory. Otherwise (4) would imply (3).

  III The first three conditions are enough to give us a countable structure whose finite substructures are exactly those in C. The last condition ensures homogeneity of the structure, as we shall see in the next theorem.

Theorem 14.4 (Fraïssé's Theorem) (i) *Let C be a class of finite $\mathcal{L}$-structures satisfying conditions* (1) – (4) *stated above. Then*

  (a) *there is a homogeneous $\mathcal{L}$-structure (called the* Fraïssé *limit) whose finite substructures are (up to isomorphism) exactly the members of C;*

  (b) *any two homogeneous $\mathcal{L}$-structures as in* (a) *are isomorphic.*

(ii) *Conversely, if M is a homogeneous $\mathcal{L}$-structure, then the class of finite $\mathcal{L}$-structures which are isomorphic to substructures of M satisfies* (1) – (4).

For a proof of the theorem, see Fraïssé (1953) or Hodges (1993, Theorem 7.1.2). We will prove the theorem in Section 14.4. But before we do that, let us look at some examples and applications to understand and appreciate the spirit and power of the theorem.

Example 14(d) Consider an empty relational system. That is, $\mathcal{L}$ is empty. Then C is the class of all finite sets. Conditions (1) and (2) are trivially satisfied. To prove (3), given $A, B \in C$, their union $A \cup B$ is also finite and hence also in C. To see (4), given $A, B_1, B_2 \in C$ and injections $f_1 : A \to B_1$, $f_2 : A \to B_2$, we can define a map g_2 mapping B_2 bijectively to a set meeting B_1 in $f_1(A)$ and such that $g_2 \circ f_2(a) = f_1(a)$ for all $a \in A$. Let g_1 be the identity map on B_1 and define D to be the finite set $B_1 \cup g_2(B_2)$. Then the set D and the maps g_1 and g_2 satisfy (4). Therefore, there exists a homogeneous structure in this language, which is a countable set with no relations.

Example 14(e) Consider a relational system $\mathcal{L}$ with one binary relation $<$, and let $\mathcal{C}$ be the class of all finite total orders. It is easy to see that conditions (1) and (2) hold.

To see that (4) holds, let B_1 and B_2 be as defined in the hypothesis of (4), so that they are two finite total orders, each having an isomorphic copy of another finite total order A in them. (Taking A to be empty will give us (3).) We shall enlarge the finite total order B_1 to get D satisfying (4). Let $A := \{a_1, a_2, \ldots, a_n\}$ with $a_1 < a_2 < \ldots < a_n$. First, for every element $a \in A$, map $f_2(a) \in B_2$ to the corresponding element $f_1(a) \in B_1$ under g_2. For an element $b_2 \in B_2$ not in $f_2(A)$, since B_2 is a total order, b_2 is either smaller than $f_2(a_1)$, bigger than $f_2(a_n)$ or lies in between $f_2(a_i)$ and $f_2(a_{i+1})$ for some i with $1 \leq i \leq n - 1$. In each case, let $g_2(b_2)$ be an element smaller than $f_1(a_1)$, bigger than $f_1(a_n)$ or some element in between $f_1(a_i)$ and $f_1(a_{i+1})$ respectively. We do this for every element of B_2, also making sure, at each step, that if $b_2 < b_2' \in B_2$, then $g_2(b_2) < g_2(b_2')$. Define D to be the set $B_1 \cup \{g_2(b_2) \mid b_2 \in B_2\}$. Then clearly D satisfies all the requirements of condition (4) with g_1 as the identity map.

Therefore by Fraïssé's Theorem there is a unique countable homogeneous structure which is the Fraïssé limit of the class $\mathcal{C}$. It follows from Cantor's Theorem that $(\mathbb{Q}, <)$ is this unique limit.

Example 14(f) Let $\mathcal{L}$ consist of a single binary relation R which is symmetric and irreflexive. Define $\mathcal{C}$ to be the class of all finite $\mathcal{L}$-structures. Then $\mathcal{C}$ is the class of all finite (undirected) graphs with no loops. It is easy to see that $\mathcal{C}$ is closed under isomorphisms and substructures. Let us denote a graph G with vertex set V and edge set E as $G = (V, E)$. Given two graphs $A = (V, E)$ and $B = (V', E')$ we define their *union* $A \cup B$ to be the graph with vertex set $V \cup V'$ and edge set $E \cup E'$. Then if A and B are finite their union is also a finite graph containing both A and B as subgraphs, so (3) holds.

In practice, when proving the amalgamation property for a class of structures, it is often easier not to work formally with embeddings, but to imagine we have two strauctures B_1 and B_2 in our class, which intersect in a structure A (the structure over which we are trying to amalgamate). We then define relations on $B_1 \cup B_2$ to obtain a structure with domain $B_1 \cup B_2$ in our class. In doing so, we are allowed to identify certain points of $B_1 \setminus B_2$ with points of $B_2 \setminus B_1$. We apply this in showing that (4) holds here.

Let A, B_1, B_2 be graphs in C such that B_1 and B_2 contain isomorphic copies of A as subgraphs. Then we can construct a graph D as follows:

(i) D has vertex set $(B_1 \setminus A) \cup B_2$ in which the isomorphic copies of A in B_1 and B_2 are identified,

(ii) two vertices in D are joined by an edge, if they are joined either in B_1 or in B_2, and

(iii) there are no edges between $B_1 \setminus A$ and $B_2 \setminus A$.

Then clearly D satisfies all the requirements of (4).

By Fraïssé's Theorem, therefore, there is a unique homogeneous countable graph Γ containing every finite graph in C as an induced subgraph. This graph is called the *(Universal) Random Graph*. Rado (1964) gave the explicit construction of the random graph using binary sequences. Erdös & Rényi (1963) gave the probabilistic construction – that is, showed that with probability 1 (given independent edge probability of $1/2$, say) any graph has the property characterising the random graph, so is isomorphic to Rado's graph (also see Erdös & Spencer 1974, Ch. 17). This graph can be shown to have many other interesting properties (cf. Truss 1985, Cameron 1990). We mention one below.

Let us say that a graph Δ has *Property* (†) if whenever U and V are finite disjoint sets of vertices in Δ, there is a vertex $x \in \Delta$ which is joined to all the vertices in U and to none of the vertices in V.

Lemma 14.5 *The random graph Γ defined above has Property* (†).

Proof: Let U and V be finite and disjoint sets of vertices. In Γ there is a subgraph of the form $U' \cup V' \cup \{x'\}$ in which x' is joined to every vertex of U' but to no vertex of V' and such that $U' \cup V'$ is isomorphic to $U \cup V$ via an isomorphism f mapping U' to U. Then by homogeneity, let us extend f to an element $\hat{f} \in \text{Aut}(\Gamma)$. Then the element $\hat{f}(x') \in \Gamma$ is joined to everything in U and to nothing in V. □

<u>EXERCISES:</u>

14(iv) Show that any two countable graphs having Property (†) are isomorphic. [Hint: Use the 'back and forth' argument.]

14(v) Form a graph with vertex set **N** where vertices m and n are adjacent if 2^n occurs in the binary expression for m or vice-versa. Show that this graph has Property (†).

14(vi) Show that there is a countable homogeneous triangle-free graph Γ (*triangle-free* means that, given any triple of vertices in Γ, it is not true that the edges between them form a triangle) such that every finite triangle-free graph is an induced subgraph.

14(vii) Show that in the triangle-free graph, the stabiliser of a point is highly transitive on the set of neighbours.

14(viii) Show that the random graph is isomorphic to

(a) the graph induced on the neighbours of a point;

(b) the graph obtained by deleting a point.

Also show that in any partition of the vertices of the random graph, one or the other part has an induced subgraph isomorphic to the random graph. [Hint: Use Lem. 14.5.]

14(ix) Show that the disjoint union of two copies of the random graph is not homogeneous, but becomes homogeneous if we add to the language a binary relation interpreted as an equivalence relation whose classes are the copies. (By 'disjoint' we mean here that there are no edges between the copies.)

The next lemma is an application of Fraïssé's Theorem.

Lemma 14.6 *For every $k \in$ **N**, there is a k-transitive but not $(k+1)$-transitive permutation group on a countably infinite set.*

Note that this situation is clearly different from the finite case, in which the only 6-transitive groups of degree n are A_n or S_n, so a

6-transitive finite group is at least $(n - 2)$-transitive.

Proof: Fix k and let $\mathcal{L}$ be the language having a single $(k + 1)$-ary relation R such that if $x_1, x_2, \ldots, x_{k+1}$ are related by R then they are distinct. Let $\mathcal{C}$ be the class of all finite $\mathcal{L}$-structures. By arguments similar to those used in Example 14(f) above, we can show that $\mathcal{C}$ satisfies conditions (1) – (4).

Thus by applying Fraïssé's Theorem, we get a homogeneous $\mathcal{L}$-structure M. Consider the group Aut (M). No automorphism of M can take a $(k+1)$-tuple which is related by R to another which is not. Therefore, Aut (M) is not $(k + 1)$-transitive. But since there is only one relation of arity $k + 1$ in M, any two k-tuples of distinct elements are unrelated. Hence any bijection between them is an isomorphism, which by homogeneity of M, can be extended to an automorphism of M. Therefore, Aut (M) is k-transitive. $\square$

14.3   Examples of homogeneous structures

The examples of the known homogeneous structures are both interesting and varied. To state an instance, Thomas (1986) has shown that one can impose a linear order on the elements of a countably infinite dimensional vector space V over a finite field of q elements, with the property that for every pair A and B of subspaces of V of the same finite dimension, there exists an order-preserving automorphism of V taking A to B. Schmerl (1979) gives a description of all countable homogeneous partially ordered sets. Note however, that the countable 2-homogeneous semilinear orders classified by Droste (1985) are $\aleph_0$-categorical partially ordered sets (cf. Ex. 14(i)) which are not homogeneous. In the rest of this section we shall discuss some other known examples of homogeneous structures. Our discussion is based mainly on Evans (1994, Section 2.2).

14(g)   Examples of homogeneous graphs:

(i) We have already seen that the universal random graph is homogeneous.

(ii) Let $\mathcal{C}$ be the class of all triangle-free graphs, as described in Exercise 14(vi). If we amalgamate two such graphs B_1 and B_2

over a common subgraph A making sure that we do not join points of $B_1 \setminus A$ to points of $B_2 \setminus A$, then the amalgamation is also triangle-free. By Fraïssé's Theorem, we can thus construct the universal triangle-free graph, which will be homogeneous.

(iii) Let K_n be the complete graph of size n, that is, a graph with n vertices in which every vertex is joined to every other vertex. For every finite n, we can define a K_n-free graph (in a way similar to the way we defined triangle-free graphs), and obtain the universal homogeneous K_n-free graph.

(iv) It is easy to verify that the disjoint union of complete graphs, all of the same size, is homogeneous (cf. Ex. 14(c)).

(v) Complements of all the examples cited above will also be homogeneous. By *complement* of a graph we mean the graph, on the same vertex set, obtained by replacing edges by non-edges and non-edges by edges.

A theorem by Lachlan & Woodrow (1980) says that these account for all the countably infinite homogeneous graphs. (There are a few finite examples also, like the pentagon, and finite versions of those mentioned in 14(g)(iii) above.)

We have already defined a directed graph in Section 5.2. A simple *digraph* or a *directed graph* is a structure with a single asymmetric, irreflexive binary relation. Henson (1972) constructed $2^{\aleph_0}$ non-isomorphic countable homogeneous digraphs. Countable homogeneous digraphs have recently been classified by Cherlin (preprint).

A special class of digraphs are the tournaments. A *tournament* is a digraph in which any two vertices are joined by a directed edge.

14(h)   Examples of homogeneous tournaments:

(i) Any totally ordered set can be easily made into a tournament by replacing $<$ by an arrow. Thus $(\mathbb{Q}, <)$, for example, is a tournament. We have already seen that it is homogeneous.

(ii) Consider the class C of all finite tournaments. Given two finite tournaments, we can amalgamate arbitrarily to get another tournament. Thus we can get a homogeneous object using the Fraïssé amalgamation technique. We call it the universal homogeneous tournament.

(iii) Take the unit circle and distribute $\aleph_0$ points densely around it, insisting that there are no antipodal points. For two such points α and β on the circle, draw an arrow from α to β if it is faster to get from α to β clockwise along the circle rather than anti-clockwise. This gives us a tournament. It can also be shown to be homogeneous.

The countably infinite homogeneous tournaments have been classified by Lachlan (1984), and a shorter proof was given by Cherlin (1988). They have shown that the three examples of homogeneous tournaments that we have described above are the only examples. (The only finite homogeneous tournaments are a single point and a 3-cycle.)

Note that the Lachlan & Woodrow Theorem implies that there are only countably many countably infinite homogeneous graphs. They are all $\aleph_0$-categorical by Theorem 14.2 and Exercise 14(ii). However, the Fraïssé amalgamation technique may be used to construct $2^{\aleph_0}$ non-isomorphic $\aleph_0$-categorical graphs. For details we refer to Droste & Macpherson (1991).

EXERCISE:

14(x) Classify the imprimitive homogeneous graphs without using the Lachlan & Woodrow Theorem. [Hint: By the D. G. Higman criterion (cf. Thm. 5.7), after replacing the graph by its complement if necessary, the graph is disconnected; by 2-homogeneity it follows that the graph is a disjoint union of complete graphs, and by transitivity, that the complete graphs must all have the same size.]

14.4 Proof of Fraïssé's Theorem

We give the statement of Fraïssé's Theorem (Theorem 14.4) once again before supplying a proof.

Fraïssé's Theorem (i) Let C be a class of finite $\mathcal{L}$-structures satisfying the conditions (1) – (4) stated in the beginning of Section 14.2. Then

(a) there is a homogeneous $\mathcal{L}$-structure (called the *Fraïssé limit*) whose finite substructures are (up to isomorphism) exactly the members of C;

(b) any two homogeneous $\mathcal{L}$-structures as in (a) are isomorphic.

(ii) Conversely, if M is a homogeneous $\mathcal{L}$-structure, then the class of finite $\mathcal{L}$-structures which are isomorphic to substructures of M satisfies (1) – (4).

Proof of (i)(a): Let C be the class of all finite $\mathcal{L}$-structures satisfying (1) – (4). We construct a homogeneous $\mathcal{L}$-structure whose finite substructures are, up to isomorphism, exactly the members of C. Define $M := \bigcup_{i \in \mathbb{N}} M_i$, where the M_i are finite structures and M_{i+1} is built up inductively from M_i as we shall see below. We have countably many steps in the construction (one for each M_i) and countably many amalgamations to perform. It suffices for us to do every possible amalgamation.

Let K be a countable set of pairs of structures (A, B) such that $A, B \in C$ and $A \subseteq B$. We can choose K so that it includes all such pairs up to isomorphism. Let $\theta : \mathbb{N} \times \mathbb{N} \to \mathbb{N}$ be a bijection such that $\theta(i, j) \geq i$ for all $i, j \in \mathbb{N}$. Start with any substructure $M_0 \in C$. At some given step, say M_k has already been defined. Then list as $(A_{kj}, B_{kj}, f_{kj})_j$ all the triples (A, B, f) where $(A, B) \in K$ and $f : A \to M_k$ is an embedding. Construct M_{k+1} applying the amalgamation property, so that if $k = \theta(i, j)$ then f_{ij} extends to an embedding of B_{ij} into M_{k+1}. Here we are amalgamating the identity mapping $\mathrm{id} : A_{ij} \to B_{ij}$ and the embedding $\mathrm{id}_{M_i} \cdot f_{ij} : A_{ij} \to M_k$ to get M_{k+1}.

This is a countable process and will eventually define M. We now claim that this M is our required structure. To see this, we first show that M has the property, call it Property $(*)$, that whenever $A, B \in \mathcal{C}$ with $A \leq B$ and $f : A \rightarrow M$ is an embedding, then f can be extended to an embedding $g : B \rightarrow M$.

Let $A \leq B$ be as defined above and $f : A \rightarrow M$ be an embedding. Since $f(A)$ is finite, there exists $j \in \mathbf{N}$ such that $f(A) \subseteq M_j$. Let the triple (A, B, f) be (A_{jl}, B_{jl}, f_{jl}). Let $k := \theta(j, l)$, so $k \geq j$. Then M_{k+1} is built by amalgamating M_k and B over $f(A)$, so f extends to some $g : B \rightarrow M_{k+1}$. Therefore M has Property $(*)$.

From the process of construction, it is clear that the finite substructures of M are exactly the members of $\mathcal{C}$ up to isomorphism. It remains to show that M is homogeneous. Suppose we have an isomorphism f between finite structures $U, V \in \mathcal{C}$. We have to find an automorphism $\hat{f}$ of M extending f. We do this using the 'back and forth' argument. At any particular step, we have finite structures $U', V' \leq M$ and an isomorphism $f' : U' \rightarrow V'$ extending f. Let $x \in M \setminus U'$. Since M has Property $(*)$ and f' is an embedding from U' to M it can be extended to an embedding f'' defined on $U' \cup \{x\}$. Thus f'' is an isomorphism from $U' \cup \{x\}$ to $V' \cup \{f''(x)\}$. In this step points are added to the domain of definition.

At the next step points are added to the range of definition, using a similar argument and Property $(*)$. Continuing in this manner, after a countable number of steps, we will have eventually extended f to an automorphism $\hat{f}$ of M.

Proof of (i)(b): Let M and N be homogeneous $\mathcal{L}$-structures with the same finite substructures, up to isomorphism. We then show, by a 'back and forth' argument, that $M \cong N$.

Since M and N are countable, we may suppose that

$$M := \{m_i \mid i \in \mathbf{N}\} \text{ and } N := \{n_j \mid j \in \mathbf{N}\}.$$

We build up the isomorphism $\phi : M \rightarrow N$ step by step.

Step 0: Define $\phi(m_0) := n_i$ where $i \in \mathbf{N}$ is the least such that $\{m_0\} \cong \{n_i\}$. Let $\mathrm{dom}(\phi)$ denote the domain of definition of the function ϕ at any given step.

Step (2i+1): Let us suppose that so far

$$\mathrm{dom}(\phi) = \{m_{k_0}, m_{k_1}, \ldots, m_{k_{2i}}\}$$

and let $j \in \mathbf{N}$ be the least j such that $m_j \notin \text{dom}(\phi)$. We will define ϕ on m_j. This step will eventually ensure that $\text{dom}(\phi) = M$.

Consider the set $\{m_{k_0}, m_{k_1}, \ldots, m_{k_{2i}}, m_j\}$. This has an isomorphic copy $\{n_{l_0}, n_{l_1}, \ldots, n_{l_{2i+1}}\}$ in N. But then

$$\{n_{l_0}, n_{l_1}, \ldots, n_{l_{2i}}\} \cong \{m_{k_0}, m_{k_1}, \ldots, m_{k_{2i}}\}$$
$$\cong \{\phi(m_{k_0}), \phi(m_{k_1}), \ldots, \phi(m_{k_{2i}})\}.$$

This isomorphism must be induced by some $g \in \text{Aut}(N)$. Define $\phi(m_j) := g(n_{l_{2i+1}})$. Then we have

$$\{m_{k_0}, m_{k_1}, \ldots, m_{k_{2i}}, m_j\} \cong \{n_{l_0}, n_{l_1}, \ldots, n_{l_{2i+1}}\}$$
$$\cong \{\phi(m_{k_0}), \ldots, \phi(m_{k_{2i}}), g(n_{l_{2i+1}})\}$$

where the second isomorphism is induced by g. Note here that all the sets under discussion here are ordered sets, and that the isomorphisms preserve order.

Step (2i+2): So far let

$$\text{range}(\phi) = \{n_{k_0}, n_{k_1}, \ldots, n_{k_{2i+1}}\}$$

and let $j \in \mathbf{N}$ be the least j such that $n_j \notin \text{range}(\phi)$. Using the same techniques as used for the odd steps, we find a suitable preimage for n_j under ϕ to eventually ensure that $\text{range}(\phi) = N$.

At the end, we will have constructed an isomorphism $\phi : M \to N$.

Proof of (ii): Let M be a homogeneous $\mathcal{L}$-structure and let $\mathcal{C}$ be the class of finite $\mathcal{L}$-structures which are isomorphic to substructures of M. We shall show that $\mathcal{C}$ satisfies conditions (1) – (4) stated earlier.

Clearly $\mathcal{C}$ is closed under isomorphism and substructures. The union of two finite substructures of M is a finite substructure of M containing both the original substructures. So the joint embedding property is also satisfied.

To prove the amalgamation property, let $A, B_1, B_2 \in \mathcal{C}$ and let $f_i : A \to B_i$, $i = 1, 2$ be embeddings. For simplicity, let us assume f_1 to be the identity map. By homogeneity of M, the isomorphism from A to $f_2(A)$ can be extended to an automorphism ϕ of M. Define $D := \phi^{-1}(B_2) \cup B_1$ and define $g_1 : B_1 \longrightarrow D$ to be the identity map and $g_2 : B_2 \longrightarrow D$ to be the restriction of the automorphism ϕ^{-1} to B_2. Then, for all $a \in A$ we have $g_2 \circ f_2(a) = \phi^{-1} \circ f_2(a) = a = g_1 \circ f_1(a)$. Only slightly more work is required to prove that the amalgamation property holds if we take f_1 to be an arbitrary embedding.

Chapter 15

The Hrushovski Construction

We have seen in the last chapter (Lemma 14.6) that for every $k \in \mathbb{N}$ we can find a k-transitive but not $(k + 1)$-transitive permutation group acting on a countably infinite set. It is natural therefore to ask whether, for every positive integer k, there is an infinite Jordan group which is k-transitive but not $(k + 1)$-transitive. In this chapter, we prove that for each $k > 1$ there is a Jordan group preserving a Steiner k-system which has this property. We use a construction technique which was first developed by Hrushovski (1993) to answer some major open questions in model theory. Our task is much simpler and our construction correspondingly shorter. See Goode (1989) or Wagner (1994) for more details. The construction given here follows Macpherson (1994, Section 5) closely.

15.1 Generalisation of Fraïssé's Theorem

The Fraïssé construction (cf. Ch. 14) can be seen as essentially a way of glueing structures together. From a different viewpoint, it can be seen as a method of piecing embeddings together. But instead of considering the class of all embeddings, we may sometimes want to consider restricted classes of embeddings and hope to obtain some version of Fraïssé's construction to work for these classes too. This section is based on Evans (1994, Section 2.1.3).

Definition 15.1 Let C be a class of finite $\mathcal{L}$-structures. We define a class of C-*embeddings* to be a collection $\mathcal{E}$ of embeddings $f : A \longrightarrow B$ (where $A, B \in C$) such that

(i) any isomorphism is in $\mathcal{E}$;

(ii) $\mathcal{E}$ is closed under composition;

(iii) if $f : A \to B$ is in $\mathcal{E}$ and $C \subseteq B$ is a substructure in C such that $f(A) \subseteq C$, then the map obtained by restricting the range of f to C is also in $\mathcal{E}$.

Suppose $\mathcal{E}$ is a class of C-embeddings. Let M be an $\mathcal{L}$-structure and A a finite substructure of M that is in C. We say that A is $\mathcal{E}$-*embedded* in M if, whenever B is a finite substructure of M which is in C and contains A, the inclusion map $A \hookrightarrow B$ is in $\mathcal{E}$. We write this as $A \leq_e M$. Note that if $f : A \to B$ is in $\mathcal{E}$ then $f(A) \leq_e B$, and if $A, B, C \in C$ and $A \leq_e B \leq_e C$ then $A \leq_e C$.

Let $\mathcal{E}$ be a class of C-embeddings. We modify the earlier definitions of the Joint Embedding Property and Amalgamation Property (cf. Sec. 14.2) as follows:

(JEP') If $B_1, B_2 \in C$ there exist $D \in C$ and embeddings $g_i : B_i \to D$ for $i = 1, 2$ such that $g_i \in \mathcal{E}$.

(AP') Suppose $A, B_1, B_2 \in C$ and $f_i : A \to B_i, i = 1, 2$ are embeddings in $\mathcal{E}$. Then there exist $D \in C$ and embeddings $g_i : B_i \to D$, for $i = 1, 2$ in $\mathcal{E}$ such that for all $a \in A$ we have $g_1 \circ f_1(a) = g_2 \circ f_2(a)$.

With these modifications, a generalised version of Fraïssé's Theorem can be stated as follows:

Theorem 15.2 (Fraïssé's Theorem Generalised) *Let C be a class of finite $\mathcal{L}$-structures and let $\mathcal{E}$ be a class of C-embeddings satisfying (JEP') and (AP'). Then there exists a countable $\mathcal{L}$-structure M with the following properties:*

(a) *the class of $\mathcal{E}$-embedded substructures is (up to isomorphism) exactly the class C;*

(b) *M is a union of finite $\mathcal{E}$-embedded substructures;*

(c) *if $A \leq_e M$ and $f : A \to B$ is in $\mathcal{E}$ then there exist $C \leq_e M$
    containing A and an isomorphism $g : B \to C$ such that we have
    $g \circ f(a) = a$ for all $a \in A$.*

*Moreover, any two $\mathcal{L}$-structures with the above properties are iso-
morphic, and any isomorphism between $\mathcal{E}$-embedded finite substructures of M extends to an automorphism of M.* $\square$

We shall apply this generalised version of Fraïssé's Theorem in
the Hrushovski construction in Section 15.3.

15.2 Building a geometry

Definition 15.3 A *pregeometry* (X, Cl) is a set X with a function
(called the *closure operator*) $\text{Cl} : \wp(X) \to \wp(X)$ satisfying the following conditions, for $A, B \in \wp(X)$ and $a, b \in X$,

(i) $A \subseteq \text{Cl}(A)$;

(ii) $A \subseteq B \Rightarrow \text{Cl}(A) \subseteq \text{Cl}(B)$;

(iii) $A \subseteq \text{Cl}(B), B \subseteq \text{Cl}(C) \Rightarrow A \subseteq \text{Cl}(C)$;

(iv) if $a \in \text{Cl}(A \cup \{b\}) \setminus \text{Cl}(A)$ then $b \in \text{Cl}(A \cup \{a\})$ *(Exchange
     Property)*;

(v) $\text{Cl}(A) = \bigcup\{\text{Cl}(F) \mid F \subseteq A,\ F \text{ finite }\}$.

The sets $\text{Cl}(A)$ for $A \in \wp(X)$ are called the *closed sets*. For $A \in \wp(X)$
the set $\text{Cl}(A)$ is called the *closure* of A.

Definition 15.4 A pregeometry is called a *geometry* if

(vi) $\text{Cl}(\emptyset) = \emptyset$; and

(vii) singleton sets are closed.

This definition makes sense in the context of vector spaces with
the closure operator meaning the linear span of the vectors in the set.

Given a pregeometry (X, Cl) and a subset A of X, we say that
A is *independent* if for every $a \in A$ we have $a \notin \text{Cl}(A \setminus \{a\})$. If
$A \subseteq B \subseteq X$ and B is closed, we say that A is a *basis* for B if A
is independent and $\text{Cl}(A) = B$. Thus a basis for a closed set is a

maximal independent subset of the set.

<u>EXERCISES</u>:

15(i) Show that conditions (i) and (iii) of Definition 15.3 together
      imply that $\mathrm{Cl}(\mathrm{Cl}(A)) = \mathrm{Cl}(A)$.

15(ii) Show that if (X, Cl) is a pregeometry and B is the closure of
       some finite subset of X, then any two bases for B have the
       same size. [Hint: Use the exchange property and imitate the
       corresponding proof for vector spaces.]

It is in fact true that if B is any closed set in X then any two
bases for B have the same cardinal number, and the reader who is
familiar with elementary transfinite techniques in set theory may find
this an interesting extension of the last exercise. The cardinality of
a base of a closed set B is known as the *dimension* of B.

Our aim in this section is to build a geometry (M, Cl_M). Towards
that end we need some more definitions.

Definition 15.5 (i) A $(k+1)$-hypergraph H is a pair (V, E) where
      V is a set and E is a collection of $(k+1)$-element subsets of
      V. We call elements of V the *vertices* or *points* of H and the
      elements of E the *(hyper)edges* of the hypergraph H. A hyper-
      graph (V, E) is *finite* if V is finite.

(ii) If $H = (V, E)$ is a finite $(k+1)$-hypergraph then we define

$$\delta(H) := |V| - |E|.$$

(iii) Given a hypergraph $H = (V, E)$, a pair (V', E'), where V' is a
       subset of V and E' is the collection of all $(k+1)$-subsets of V'
       that also belong to E, is called a *subhypergraph* of H. We write
       $H' \leq H$ if H' is a subhypergraph of H.

Let $\mathcal{H}_{k+1}$ be the class of all $(k+1)$-hypergraphs and let $\mathcal{A}_k$ be
defined to be the subset

$$\{H \in \mathcal{H}_{k+1} \mid (K \leq H, \text{ and } K \text{ finite}) \Rightarrow (\delta(K) \geq \min(|V_K|, k))\}.$$

where V_K is the vertex set of K. So, finite elements of $\mathcal{A}_k$ have more vertices than edges and this property is inherited by substructures. In particular, if $H \in \mathcal{A}_k$ is finite, then $\delta(H) > 0$.

Definition 15.6 If $A, M \in \mathcal{A}_k$ with A finite and $A \leq M$, we write $A \leq_{we} M$ and say that A is *well-embedded* in M, if for any finite B with $A \leq B \leq M$ we have $\delta(A) \leq \delta(B)$.

The idea behind this definition is that as we add points of M to A the number of vertices increases at least as fast as the number of edges.

Lemma 15.7 *If $A, M \in \mathcal{A}_k$ with A finite and $A \leq M$, then there exists finite $B \in \mathcal{A}_k$ with $A \leq B \leq_{we} M$.*

Proof: Choose B with $A \leq B \leq M$ and $\delta(B)$ as small as possible. Then it follows that such a hypergraph B is well-embedded in M. $\square$

Lemma 15.8 *If A and B are finite such that $A \leq_{we} B$ and $B \leq_{we} M$ then $A \leq_{we} M$.*

Proof: Suppose there exists $C \in \mathcal{A}_k$ such that $A \leq C \leq M$ and $\delta(C) < \delta(A)$. Clearly, $C \not\leq B$ and $B \not\leq C$. Also, $\delta(C \cap B) \geq \delta(A)$, so adjoining $C \setminus B$ to $C \cap B$ adjoins more edges than vertices. Hence adjoining $C \setminus B$ to B adds more edges than vertices. But in that case we have $\delta(C \cup B) < \delta(B)$, a contradiction. $\square$

Lemma 15.9 *If $A, B, M \in \mathcal{A}_k$ with A, B finite and well-embedded in M then $A \cap B$ is also well-embedded in M.*

Proof: Let $A \cap B \leq C \leq M$ for finite C. We want to show that $\delta(C) \geq \delta(A \cap B)$. Now, since $A \leq_{we} M$ we have $\delta(C \cup A) \geq \delta(A)$. This means that adding $C \setminus A$ to A adds at least as many vertices as edges. This in turn means that adding $C \setminus A$ to $C \cap A$ adds at least as many vertices as edges. But this implies $\delta(C) \geq \delta(C \cap A)$.

Similarly, since $B \leq_{we} M$ we have $\delta((C \cap A) \cup B) \geq \delta(B)$. Applying the same argument as above we get

$$\delta(C \cap A) \geq \delta(C \cap A \cap B) = \delta(A \cap B).$$

Therefore, $\delta(C) \geq \delta(A \cap B)$ which is what we wanted to prove. $\square$

The following lemma is an immediate corollary of Lemmas 15.7 and 15.9.

Lemma 15.10 *If $A, M \in \mathcal{A}_k$ with A finite and $A \leq M$ then there is a unique smallest finite $B \in \mathcal{A}_k$ with $A \leq B \leq_{we} M$.* □

Define B as in the last lemma to be the *envelope* of A in M and denote it by $\mathrm{Env}_M(A)$ and put

$$d_M(A) := \delta(B) = \delta(\mathrm{Env}_M(A)).$$

Clearly we always have $d_M(A) \leq \delta(A)$. Note that the envelope is defined only for finite $A \in \mathcal{A}_k$. We have deliberately diverged from the model-theoretic terminology set up by Hrushovski (1993) in order to avoid confusion with different notions of closure. In the survey by Macpherson (1994), the 'envelope' is called 'self-sufficient closure'. and $A \leq_{we} M$ is read A is self-sufficient in M. We use the envelope to define the concept of closure needed in a geometry.

Definition 15.11 For any $M \in \mathcal{A}_k$ and for finite $A \leq M$. we define the *closure* of A in M to be

$$\mathrm{Cl}_M(A) := \{x \in M \mid d_M(A \cup \{x\}) = d_M(A)\}.$$

For $N, M \in \mathcal{A}_k$ and $N \leq M$, we put

$$\mathrm{Cl}_M(N) := \bigcup \{\mathrm{Cl}_M(A) \mid A \leq N, A \text{ finite }\}.$$

EXERCISE:

15(iii) (a) Show that if A and B are finite with $A \subseteq B \subset M$ then
$d_M(A) \leq d_M(B)$.

(b) Show that if A and B are finite with $A \subseteq B \subseteq \mathrm{Cl}_M(A)$,
then $d_M(A) = d_M(B)$.

Since $\delta(\mathrm{Env}_M(A)) = d_M(A)$, it follows that if $x \in \mathrm{Env}_M(A)$ then $d_M(A) = d_M(A \cup \{x\})$. Therefore $A \leq \mathrm{Env}_M(A) \leq \mathrm{Cl}_M(A) \leq M$. Also note that for finite A, the envelope is always finite while the closure is often infinite, as the following exercise will demonstrate.

EXERCISE:

15(iv) Let M be the set of natural numbers, A be the set $\{0, \ldots, k\}$, and suppose that the edges on M are the $(k+1)$-sets of the form $\{l, l+1, \ldots, l+k+1\}$ (where $l \in \mathbb{N}$) except for A. Show that the closure of A in M is M.

Lemma 15.12 If $A, M \in \mathcal{A}_k$ with A finite and such that $A \leq M$, and if $x \in M \setminus A$, then $x \in \mathrm{Cl}_M(A)$ if and only if some edge of $\mathrm{Cl}_M(A) \cup \{x\}$ meets x.

Proof: Suppose $x \in \mathrm{Cl}_M(A)$, so $d_M(A) = d_M(A \cup \{x\})$. Then $\mathrm{Env}_M(A) \subseteq E_x := \mathrm{Env}_M(A \cup \{x\}) \subseteq \mathrm{Cl}_M(A)$, and we must have $\delta(E_x) = \delta(\mathrm{Env}_M(A))$. Now some edge of E_x meets x, since otherwise $\delta(E_x \setminus \{x\}) < \delta(\mathrm{Env}_M(A))$, contrary to the definition of $\mathrm{Env}_M(A)$. The converse is an exercise. $\square$

Lemma 15.13 If $M \in \mathcal{A}_k$ then (M, Cl_M) is a geometry.

Proof: (i) Clearly, $A \subseteq \mathrm{Cl}_M(A)$ for all $A \subseteq M$.

(ii) Let $A \subseteq B$. Then by (i) $A \subseteq B \subseteq \mathrm{Cl}_M(B)$. Suppose some $x \in \mathrm{Cl}_M(A)$ does not belong to $\mathrm{Cl}_M(B)$. Then from Lemma 15.12 it follows that no edge of $\mathrm{Cl}_M(B) \cup \{x\}$ meets x. This in turn implies that no edge of $\mathrm{Cl}_M(A) \cup \{x\}$ meets x. Applying Lemma 15.12 once again we get that $x \notin \mathrm{Cl}_M(A)$, a contradiction.

(iii) Let $A \subseteq \mathrm{Cl}_M(B)$ and $B \subseteq \mathrm{Cl}_M(C)$. Then from (i) and (ii) it follows that $A \subseteq \mathrm{Cl}_M(\mathrm{Cl}_M(C))$. To show $Cl_M(Cl_M(C) \subseteq Cl_M(C)$, let $x \in Cl_M(Cl_M(C)$. Then $x \in Cl_M(D)$ for some finite $D \subseteq Cl_M(C)$, and by Exercise 15(iii)(a) we may assume that $C \subseteq D$. Hence $d_M(D \cup \{x\}) = d(D)$. Also, by Exercise 15(iii)(b), it follows that $d_M(D) = d_M(C)$, so that $d(D \cup \{x\}) = d(C)$. Clearly

$$d_M(C) \leq d_M(C \cup \{x\}) \leq d_M(D \cup \{x\})$$

(by Ex. 15(iii)(a)), whence $d_M(C) = d_M(C \cup \{x\})$, so $x \in Cl_M(C)$, as required.

(iv) For the exchange property, suppose that $A \subseteq M$, that $b, c \in M$ and that $c \in \mathrm{Cl}_M(A \cup \{b\}) \setminus \mathrm{Cl}_M(A)$. We may suppose that A is finite. Then $d_M(A \cup \{b, c\}) = d_M(A \cup \{b\})$ and $d_M(A \cup \{c\}) = d_M(A) + 1$. It follows that $d_M(A \cup \{b\}) = d_M(A) + 1$. For otherwise, it would be the case that $d_M(A \cup \{b\}) = d_M(A)$, so that

$$d_M(A \cup \{c\}) \leq d_M(A \cup \{b, c\}) = d_M(A \cup \{b\}) = d_M(A),$$

a contradiction. Therefore, we have $d_M(A \cup \{b, c\}) = d_M(A \cup \{c\})$. which implies that $b \in \mathrm{Cl}_M(A \cup \{c\})$.

(v) This follows directly from the definition of Cl_M.

(vi) Clearly $\mathrm{Cl}_M(\emptyset) = \emptyset$ as $\delta(\emptyset) = 0$.

(vii) Finally, singleton subsets A are closed. This is because $\delta(A) = 1$ and if B has at least 2 points, then $\delta(B) \geq 2$. □

15.3 Extending the geometry

We shall use the geometry constructed in the last section to build a geometry of dimension $\aleph_0$. For this we use the generalised version of Fraïssé's amalgamation technique as discussed in Section 15.1. But first we need to define the notion of a free amalgamation.

Definition 15.14 If $A, B, C, D \in \mathcal{A}_k$ with $A \leq B$, $A \leq C$ and $D = B \cup C$ then we say that D is a *free amalgamation* of B and C over A, and denote it by $B \bigoplus_A C$, if $B \cap C = A$ and no edge of D meets both $B \setminus A$ and $C \setminus A$.

In the following lemma, it is shown that a version of amalgamation is possible.

Lemma 15.15 *Let $A, B, C \in \mathcal{A}_k$ be finite with $A \leq_{we} B$, $A \leq_{we} C$. Put $D := B \bigoplus_A C$. Then $D \in \mathcal{A}_k$ and $B \leq_{we} D$, $C \leq_{we} D$.*

Proof: To prove that $B \leq_{we} D$, suppose there exists some finite $E \in \mathcal{A}_k$ with $B \leq E \leq D$ such that $\delta(E) < \delta(B)$. Then adjoining $E \setminus B$ to B adds more edges than vertices. Since no edge of D meets both $B \setminus A$ and $C \setminus A$, adjoining $E \setminus B$ to A adds more edges than vertices. But this contradicts the fact that $A \leq_{we} C$. Similarly we can show that $C \leq_{we} D$.

A similar argument proves that $D \in \mathcal{A}_k$. □

The last lemma gives us all the ingredients necessary to apply the generalised version of Fraïssé's Theorem as stated in Section 15.1. We restate Theorem 15.2 in the language of well-embeddings. We may regard hypergraphs as structures in a language with a single relation symbol of arity $k+1$. The relation will be symmetric in its arguments, and will hold precisely of $(k + 1)$-tuples which are orderings of edges.

Theorem 15.16 *There is a unique (up to isomorphism) countably infinite structure $M \in \mathcal{A}_k$ such that*

(a) M contains a copy of every finite $A \in \mathcal{A}_k$, and that copy of A is well-embedded in M;

(b) M is a union of finite $A \in \mathcal{A}_k$ such that $A \leq_{we} M$;

(c) whenever $A, B \in \mathcal{A}_k$ are finite and $f : A \to B$ and $g : A \to M$ are embeddings such that $f(A) \leq_{we} B$ and $g(A) \leq_{we} M$, there is an embedding $h : B \to M$ such that $h \circ f = g$ and $h(B) \leq_{we} M$.

Moreover, any two well-embedded structures with the above properties are isomorphic, and any isomorphism between well-embedded finite substructures of M extends to an automorphism of M.

Proof: We build M as a union of a countable chain of finite substructures. As in the proof of Fraïssé's Theorem (cf. Sec. 14.4) we make a countable list K of pairs of structures (A, B) such that $A, B \in \mathcal{A}_k$ and $A \leq_{we} B$. We then build M_{i+1} from M_i inductively, by amalgamating M_i and B over A for the next pair (A, B) in the list K, exactly as in the proof of Fraïssé's Theorem, in such a way that both M_i and B are well-embedded in M_{i+1}.

The first three properties follow from the definition of M. There are countably many steps in the chain, which ensures that the construction is possible. The last two assertions of the theorem are proved by 'back and forth' arguments, much as in the proof of Fraïssé's Theorem. $\square$

The last theorem gives us a geometry (M, Cl_M) of countably infinite dimension which we shall use in the proof of Theorem 15.18. However, M is not $\aleph_0$-categorical as we shall see in the following lemma.

Lemma 15.17 *M is not $\aleph_0$-categorical.*

Sketch Proof: We find $(k + 1)$-sets with arbitrarily large finite envelopes. Since any finite structure determines its envelope, two structures with non-isomorphic envelopes will lie in different orbits, so all these $(k + 1)$-sets will lie in different orbits.

To construct a $(k + 1)$-set X_m with an envelope E_m of size m where $m > 2k + 1$, start with a $(k + 1)$-set $X := \{x_1, .., x_{k+1}\}$, which is a non-edge. Let $E_m := \{x_1, \ldots, x_{k+1}, y_1, \ldots, y_n\}$, where we have

$n = m - (k+1)$, and the sets $\{x_2, \ldots, x_{k+1}, y_1\}$, $\{x_3, \ldots, x_{k+1}, y_1, y_2\}$, $\ldots, \{y_{n-(k-1)}, \ldots, y_n, x_1\}$ are all edges. It is easy to check that X_m is well-embedded in E_m, and $\delta(E_m) = k$, whereas for any H with $X_m \leq H < E_m$ we have $\delta(H) \geq k + 1$. There is a copy of X_m in M, and E_m will be its envelope in M (by Lemma 15.9 and the above observation on H), as required. □

Theorem 15.18 *For every* $k \in \mathbf{N}$ *with* $k > 1$, *there is a group preserving a Steiner* k-*system which is* k-*transitive but not* $(k + 1)$-*transitive.*

Proof: We have already constructed a geometry $(M, \mathrm{Cl}\,_M)$ of countably infinite dimension. Set $G := \mathrm{Aut}\,(M)$, that is, G is the automorphism group of M as a hypergraph. Then G preserves the closure operator (and therefore maps closed sets to closed sets) as it is defined in terms of the hypergraph structure on M.

We define a Steiner k-system with points the elements of M and blocks the sets of the form $\mathrm{Cl}\,_M(A)$ with $|A| = k$. That is, the closed sets of dimension k are the blocks of the Steiner k-system on M.

Any k-set A is well-embedded (since $\delta(A) = k$ if $|A| = k$ and δ cannot take a smaller value than k on a larger set). The last part of Theorem 15.16 then ensures that G has the level of homogeneity required to extend any bijection between two sets of size k to an automorphism of M. Therefore G is k-transitive. But G is not $(k+1)$-transitive as it is the automorphism group of a $(k + 1)$-hypergraph (and therefore edges and non-edges of M lie on different orbits of G).

Next, if $A := \{a_1, a_2, \ldots, a_{k+1}\}$ is an edge then $\delta(A) = k$, so that $a_{k+1} \in \mathrm{Cl}\,_M(\{a_1, a_2, \ldots, a_k\})$, so by k-transitivity, the blocks, which are the closures of k-sets, have more than k points. Note also that any two edges are well-embedded, and hence lie in the same orbit. The fact that k points lie on a unique block follows from the closure axioms.

It only remains to show that M is not the closure of a k-set (to ensure that there is more than one block). Choose A to be a k-set in M and B to be a $(k + 1)$-set containing A which is a non-edge. We then have $A \leq_{we} B$, so that there is some well-embedded copy B' of B in M containing A. Then $d_M(B') = \delta(B') = k + 1$, so that $B' \not\subseteq \mathrm{Cl}\,_M(A)$. □

We shall end this chapter by showing that G, the group preserving the Steiner k-system we have defined in the last theorem, is a Jordan group. We call subsets of M which are closed in M *subspaces* of M. Let us first prove the following partial converse of Lemma 15.15.

Lemma 15.19 *Let $M \in \mathcal{A}_k$ and let X be a finite dimensional subspace of M. Also let B be finite, $B \leq_{we} M$ and set $A := B \cap X$ and suppose that $d_M(A) = d_M(X)$. Then X and B form a free amalgamation over A.*

Proof: Let C be finite with $A \leq C \leq X$ and put $C' := \mathrm{Env}_M(C)$. Then $C' \leq X$, and by Lemma 15.9, we have $A = B \cap C' \leq_{we} M$. Hence $\delta(C') = d(C') = d(A) = \delta(A)$. If there was a hyperedge of $B \cup C'$ meeting both $B \backslash A$ and $C' \backslash A$ then, since adding $C' \backslash A$ to A adds the same number of edges as vertices, adding $C' \backslash A$ to B would add more edges than vertices. Hence we would have, $\delta(B \cup C') < \delta(B)$, a contradiction to the fact that $B \leq_{we} M$. $\square$

The main point of this chapter lies in the following theorem.

Theorem 15.20 *The complement in M of any finite dimensional closed set is a proper Jordan set for G.*

Proof: Let X be a finite dimensional subspace of M and let a and b be elements of $M \setminus X$. We want to find $g \in G_{(X)}$ which maps a to b. The construction is by a 'back and forth' argument, but it is a little delicate, since the map g which we build must be the identity on the whole of X. In previous 'back and forth' arguments, at any given stage the map has been determined only on a finite set, but here it is determined from the beginning on the infinite set X.

For any finite $A \leq X$, we have

$$d_M(A \cup \{a\}) = d_M(A) + 1 = d_M(A \cup \{b\}).$$

Hence no edge of M lies in $X \cup \{a\}$ and contains a, or in $X \cup \{b\}$ and contains b. Thus the function f defined on $X \cup \{a\}$ fixing X pointwise and taking a to b is a k-hypergraph isomorphism. Note that for any finite $A \leq_{we} X$, since $a \notin \mathrm{Cl}_M(A)$, we must have $A \cup \{a\} \leq_{we} M$. Similarly, $A \cup \{b\} \leq_{we} M$.

We shall build the element $g \in \text{Aut}\,(M)$ fixing X pointwise and taking a to b by a 'back and forth' argument. At the n-th stage we will have built partial automorphisms $g_0 \subseteq g_1 \subseteq \ldots \subseteq g_n$, where g_i is defined on the set $X \cup A_i$ with $A_0 \subseteq A_1 \subseteq \ldots \subseteq A_n$, A_n finite and well-embedded in M and such that $\text{Cl}_M(A_n \cap X) = X$. From Lemma 15.19 it follows that for any finite B with $A_n \cap X \le B \le X$, the set $B \cup A_n$ forms a free amalgamation over $B \cap A_n$.

At Step 0, we may choose g_0 to be the identity mapping on X and A_0 to be some finite set containing a basis for X.

At Step 1, we may choose g_1 to be the element f already defined and A_1 to be $A_0 \cup \{a\}$.

Suppose after Step n we have defined g_n, A_n satisfying the requirements above and let $x \in M \setminus (X \cup A_n)$. Then we want to extend g_n to g_{n+1} in such a way that $x \in \text{dom}(g_{n+1})$. This is the 'forth' step. In the next 'back' step we similarly add elements to the range of definition. Set $A_{n+1} := \text{Env}_M(A_n \cup \{x\})$. Then by the last lemma it follows that for any finite B with $A_{n+1} \cap X \le B \le X$ we have that B and A_{n+1} form a free amalgamation over $A_{n+1} \cap X$.

We then claim that $A_n \cup (X \cap A_{n+1}) \le_{we} M$. To see this, first note that by Lemma 15.9, both $(X \cap A_{n+1}), (X \cap A_n) \le_{we} M$. Hence $\delta(X \cap A_{n+1}), \delta(X \cap A_n) \le \delta(X)$. But since A_n contains a basis for X it is also true that $\delta(X \cap A_{n+1}), \delta(X \cap A_n) \ge \delta(X)$. Hence it follows that $\delta(X \cap A_{n+1}) = \delta(X \cap A_n) = \delta(X)$. But since A_n and $X \cap A_{n+1}$ form a free amalgamation over their intersection, it follows that $\delta(A_n \cup (X \cap A_{n+1})) = \delta(A_n)$. Then, since A_n is well-embedded in M, it is also true that $A_n \cup (X \cap A_{n+1}) \le_{we} M$.

Let $\hat{g}_n$ be the restriction of g_n to $A_n \cup (X \cap A_{n+1})$. Then by 'homogeneity' of M and the fact that $A_n \cup (X \cap A_{n+1}) \le_{we} M$, we can extend $\hat{g}_n$ to some $h_n \in G$. Let k_n be the restriction of h_n to $X \cup A_{n+1}$. Since k_n fixes pointwise a basis of X, it fixes X setwise and hence induces an automorphism l_n of X. Also, k_n fixes $X \cap A_{n+1}$ pointwise. Since X and A_{n+1} form a free amalgamation over their intersection, the mapping $\hat{l}_n$ fixing A_{n+1} pointwise and inducing l_n^{-1} on X is a hypergraph isomorphism. Set $g_{n+1} := k_n \hat{l}_n^{-1}$. Then $x \in \text{dom}(g_{n+1})$ and this element g_{n+1} satisfies all the conditions of the inductive hypothesis. This completes the proof of the theorem. $\square$

Chapter 16

Applications and Open Questions

In this chapter we will state some applications of the material contained in this book, most specifically of the classification theorem for infinite primitive Jordan groups. We have already seen many applications of the Fraïssé amalgamation technique in Chapters 14 and 15. We will also state some questions which are of current research interest and indicate what is known and where to look for more information on these topics.

16.1 Bounded groups

Recall that the *support*, supp(g), of a permutation g is the set of elements moved by g. Recall also that Wielandt's Theorem (Theorem 6.8) states that every primitive permutation group on an infinite set Ω which contains a non-identity element of finite support must contain the finitary alternating group.

Neumann (1975, 1976) has developed an elegant theory for finitary permutation groups, that is, permutation groups of infinite degree all of whose elements have finite support. This theory has had many applications, for example, in recent work by J. Hall and others on finitary linear groups. Cameron (1996) has recently come up with the 'dual' notion of cofinitary permutation groups.

171

To discuss the uncountable analogues of Wielandt's theorem, we make the following definitions.

Definition 16.1 (i) If $G \leq \text{Sym}(\Omega)$ and m is an infinite cardinal with $m < |\Omega|$, then G is said to be *m-bounded* if for all $g \in G$,

$$|\text{supp}(g)| \leq m.$$

(ii) If $G \leq \text{Sym}(\Omega)$, then the *minimal degree* of G is

$$\min\{|\text{supp}(g)| \mid g \in G, g \neq 1\}.$$

A study of infinite bounded permutation groups can be found in Adeleke & Neumann (1996b). We state a couple of major results.

Theorem 16.2 (Adeleke & Neumann 1996b) *Let G be a primitive m-bounded permutation group on Ω where m is an infinite cardinal. Then*

(i) *G is a Jordan group with proper primitive Jordan sets, and*

(ii) *G is highly transitive or preserves a linear order, semilinear order or a C-relation.*

Theorem 16.3 *Let G be a primitive permutation group on an infinite set Ω and with minimal degree less than $|\Omega|$. Then either the conclusion of Theorem 16.2 (ii) holds or G is 2-transitive and preserves a linear betweenness relation.*

16.2   Cycle types

For a permutation g on a set Ω we have already defined the cycle type of g in Section 2.3. One general problem is to identify which cycle types (on a countable set) are realisable by a primitive permutation group of countably infinite degree which is not highly transitive.

In one direction, the best results known are due to Truss (1985), in which there is a classification of the cycle types which are realised by automorphisms of the k-coloured random graph (which is the universal homogeneous object in a language with k symmetric and irreflexive and mutually exclusive binary relations for some $k \leq \aleph_0$).

It can be shown that any cycle type which includes infinitely many infinite cycles is realised by the automorphism group of the random graph. We define cycle-by-cycle a graph structure on the set on which the permutation acts, ensuring as we go along that the permutation is an automorphism. There will be no edges within a cycle. Given any two finite disjoint sets U and V of vertices, pick a vertex in a new infinite cycle and specify that it is joined to all members of U and to no members of V.

Macpherson & Praeger (1995) have worked on this question from a slightly different angle.

Definition 16.4 A cycle type σ is *rare* if whenever G is a primitive permutation group on an infinite set Ω, and some element of G has cycle type σ, then G is highly transitive.

PROBLEM 1. Classify rare cycle types.

By Wielandt's Theorem, non-identity finitary permutations are rare. We also have the following theorem.

Theorem 16.5 (Macpherson & Praeger 1995) *If a cycle type σ consists of a single infinite cycle, infinitely many fixed points and a finite number (greater than zero) of non-trivial finite cycles, then it is rare.*

Comment on the proof: The points in the infinite cycle are very close to being a Jordan set, provided one can handle the finite cycles. So a Jordan set is built out of the support of the permutation with the cycle type σ. If (G, Ω) is a primitive permutation group containing an element of cycle type σ, then one shows, by induction on the degree of transitivity, that (G, Ω) is contained in a Jordan group of the same degree of transitivity. Then the classification of the primitive Jordan groups is used to prove the claim.

But what about other cycle types, for example, those with two infinite cycles instead of one? Or more generally, we can ask the following question.

QUESTION 2. If we allow $k \geq 1$ infinite cycles in the cycle type described in Theorem 16.5, then does the theorem still hold?

Let σ be the cycle type with just one infinite cycle and infinitely many fixed points. Any primitive group realising σ is a Jordan group, but σ is not rare. Such a cycle type is realised by certain 2-transitive groups which preserve some C-relations.

If we look at the cycle type with two infinite cycles and infinitely many fixed points (and no other finite cycles) then it is not rare, as it can be shown to be realised in the automorphism group of a certain C-relation. But it is conjectured that it is 'almost rare' in the sense that the only groups that realise it are Jordan groups.

All these observations led Macpherson and Praeger to make the following conjectures:

CONJECTURE A: Any primitive but not highly transitive permutation group with a non-identity element having finitely many non-trivial finite cycles, infinitely many fixed points, and a finite number of infinite cycles is contained in a Jordan group which is not highly transitive.

CONJECTURE B: If a cycle type consists of finitely many infinite cycles, infinitely many fixed points and a finite number (greater than zero) of non-trivial finite cycles, then it is rare.

The questions in this section are on the possible orbit structure of cyclic subgroups of a primitive but not highly transitive permutation group. Similar questions can also be asked for the orbit structure of *arbitrary* subgroups.

16.3   Strictly minimal sets

Definition 16.6 A *strictly minimal set* is a first order structure M such that

(i) $G := \operatorname{Aut}(M)$ is primitive and oligomorphic, and

(ii) for every finite $A \subseteq M$, the group $G_{(A)}$ has a cofinite orbit on M.

Suppose that M is a strictly minimal set, and let $A \subset M$ be finite. Then $G_{(A)}$ has a cofinite orbit X on M, and as $|G_{(A)} : G_{(M \backslash X)}|$ is finite and each orbit of $G_{(M\backslash X)}$ is finite or cofinite, $G_{(M\backslash X)}$ is transitive on X, so X is a Jordan set for G. Thus, each finite subset of M is contained in a finite Jordan complement.

Strictly minimal sets play a major role in model theory. They are the building blocks for *totally categorical structures*, that is, infinite first-order structures M such that the first order sentences true of M have a unique model up to isomorphism of each infinite cardinality. The following theorem classifies strictly minimal sets.

Theorem 16.7 (Cherlin *et al* 1985, Zil'ber 1979, 1984) *Let M be strictly minimal. Then one of the following holds:*

(i) $\operatorname{Aut}(M) = \operatorname{Sym}(M)$;

(ii) $M = \operatorname{PG}(V)$, *where V is an infinite-dimensional vector space over a finite field, and $\operatorname{PGL}(V) \leq \operatorname{Aut}(M) \leq \operatorname{P\Gamma L}(V)$;*

(iii) $M = \operatorname{AG}(V)$ *and $\operatorname{AGL}(V) \leq \operatorname{Aut}(M) \leq \operatorname{A\Gamma L}(V)$, where V is as in (ii) above.*

Remarks:

I The groups $\operatorname{PG}(V)$ and $\operatorname{AG}(V)$ denote the projective and affine space respectively, and $\operatorname{P\Gamma L}(V)$ and $\operatorname{A\Gamma L}(V)$ denote the projective and affine semilinear groups respectively, which we have defined in Section 7.4.

II The field F is necessarily finite here because if $\dim(V) \geq 2$ and F is infinite then the associated projective and affine groups are not oligomorphic. To see this, let u and v be linearly independent elements of V. The quadruples $(\langle u \rangle, \langle v \rangle, \langle u + v \rangle, \langle u + \alpha v \rangle)$ and $(\langle u \rangle, \langle v \rangle, \langle u + v \rangle, \langle u + \beta v \rangle)$ lie in different $\operatorname{PGL}(V)$-orbits if $\alpha \neq \beta$. Similarly, the triples $(0, u, \alpha u)$ and $(0, u, \beta u)$ lie in different $\operatorname{AGL}(V)$-orbits if $\alpha \neq \beta$.

III The proof by Cherlin (and independently by, Mills) is given in Cherlin *et al* (1985). It uses the classification of finite simple groups and the fact that M contains many cofinite Jordan sets for G (see Section 10.4), and that finite primitive Jordan groups are 2-transitive, and are hence classified. A geometrical proof was given by Zil'ber (1979, 1984). A combinatorial proof was given by Evans (1986). A different proof was given by Hrushovski (1992).

Neumann (1985) looked at strictly minimal sets independently and for completely different reasons and came up with essentially the same answer. A *moiety* of a countably infinite set Ω is a subset Σ of Ω such that both Σ and $\Omega \setminus \Sigma$ are infinite. Since Ω has only countably many finite subsets and only countably many cofinite subsets there are $2^{\aleph_0}$ moieties on Ω. Moreover, if G is a group of permutations of Ω then G permutes the moieties among themselves. With these definitions, we have the following theorem.

Theorem 16.8 (Neumann 1985) *Suppose G is a primitive permutation group on a countably infinite set Ω, with no countable orbits on moieties. Then G is 2-transitive. Furthermore, either G is highly transitive, or G acts on a projective or affine space over a finite field.*

The link between the two theorems lies in the fact that both the group G defined in Definition 16.6 as well as the group G in Theorem 16.8 have the property that the stabiliser in G of any finite tuple has just one infinite orbit, which contains all but finitely many points; and this orbit is a cofinite Jordan set. Theorem 16.8 is then a consequence of the classification theorem of primitive groups with cofinite Jordan sets (Theorem 10.16).

16.4   'Back and forth' arguments

In Section 9.1 we have seen two proofs of Cantor's Theorem, one based on 'going forth' and the other on going 'back and forth'. If M is an $\aleph_0$-categorical structure (perhaps also homogeneous in a finite relational language) we say that *forth suffices for M* if the following holds: whenever $\{m_i \mid i \in \mathbf{N}\}$, and $\{n_i \mid i \in \mathbf{N}\}$ are enumerations of M, the 'forth' part of the 'back and forth' argument yields an automorphism. By the 'forth' part, we mean the construction in which, having built a partial automorphism f_k at the k-th step, at the next step we extend f_k by mapping m_{k+1} to a suitable element of least index in the n_i enumeration. We have seen an example (in Section 9.2) in which 'going forth' does not suffice.

PROBLEM 3. Characterise all structures for which 'going forth' suffices.

McLeish (1994) studied this problem for first-order homogeneous structures. The question whether 'going forth' suffices in proofs of isomorphisms of $\aleph_0$-categorical structures leads to questions about Jordan groups. For more on this subject see Cameron (1990, pp. 124–129) and also Neumann (1996).

16.5 Homogeneous structures

We have already seen classifications of some classes of homogeneous first order structures in Chapter 14. One can try to arrive at similar classifications of other structures.

PROBLEM 4. Classify homogeneous

(a) structures with a single ternary relation;

(b) bipartite graphs, namely graphs in which the vertices can be coloured with two colours in such a way that no two vertices with the same colour are adjacent;

(c) graphs in which edges are coloured with a fixed number, greater than one, of colours.

These questions are very important and difficult. Despite the classification theorems, homogeneous structures are still very mysterious.

There are other more general questions that one might ask. The following question was publicised by Cherlin.

QUESTION 5. It is known that a first order sentence true for the universal random graph is also true for some finite subgraph of it. Is the corresponding theorem true when we replace the universal random graph by the universal triangle-free graph, or the K_n-free graph?

16.6 Structure of Jordan groups

There are various open questions relating to the structure of Jordan groups.

Consider closed Jordan groups on a countable set. That is, groups of the form Aut ($\mathcal{M}$) for some countable first-order structure

$$\mathcal{M} := (M, R_1, R_2, \ldots,).$$

QUESTION 6. When is it possible to expand the structure on M (that is, add more structure to the domain M) and still be left with a Jordan group?

QUESTION 7. Can one classify all such expansions?

For a specific instance, given a C-relation C on a set M, consider the primitive Jordan group Aut($\mathcal{M}$), where $\mathcal{M} := (M, C)$. In his study Cameron (1987) has come up with quite a few interesting examples of extra structures that can be imposed on $\mathcal{M}$. We ask for a complete catalogue.

PROBLEM 8. Describe *all* ways in which a C-relation can be combined with other structures, to obtain an object with a Jordan automorphism group.

Example A: Draw a semilinear order in the plane, and let M be a dense set of maximal chains, with the induced C-relation. Endow M in addition with a dense total order suggested by the picture, in which each cone determines a convex set.

Now consider the group $G := \mathrm{Aut}(M, C, <)$. It is easy to see that cones determine intervals (that is, the set of chains in any cone is a convex set). Under certain homogeneity conditions, the group G will be primitive (indeed 2-homogeneous) and cones will be Jordan sets. We present a concrete version of this example in the following problem.

PROBLEM 9. Consider sequences of rationals of finite support, indexed by $\mathbb{Q}$. Endow this set with the natural C-relation, as in the construction of 12.6, and a lexicographic total order. Investigate its automorphism group.

Example B: Consider the same object (M, C) as in Example A, but colour all the internal nodes of the underlying semilinear order red or green densely. Define a symmetric binary relation R to hold between two maximal chains α and β if and only if they meet at a red node.

QUESTION 10. Form the structure (M, C, R). What can be said about its automorphism group?

Cones are clearly Jordan sets here.

It is also possible to embed a non-linear betweenness relation in the plane, so that the set of directions (or a dense subset of it) carries a D-relation with an induced circular order or separation relation, with a Jordan (and oligomorphic) automorphism group. This is analogous to Example A.

PROBLEM 11. Classify all primitive Jordan subgroups of the projective and affine linear and semilinear groups.

This problem is posed, *albeit* in a different language, in Neumann & Praeger (1995).

We have already commented in Chapter 13 that the limit structures obtained in the classification theorem of infinite primitive Jordan groups are very complicated objects and are not very well understood.

PROBLEM 12. Give a satisfactory description of limits of betweenness relations, D-relations and Steiner systems.

PROBLEM 13. Describe all classes of oligomorphic primitive closed Jordan groups.

All the familiar relational structures admitting Jordan groups, yield oligomorphic examples, and a complete classification seems quite feasible. But it is not known if there are any examples of oligomorphic Jordan groups which preserve any of the limit structures. Another question that can be asked is, given $k > 3$, is there an oligomorphic example of a Jordan group which is k-transitive but not $(k + 1)$-transitive?

We have seen some examples of highly transitive groups in Chapter 11. The following problem asks for a systematic investigation.

PROBLEM 14. Investigate the highly transitive Jordan groups.

16.7   Generalisations of Jordan groups

Some parts of the classification theorem (of primitive Jordan groups) go through without the full assumption that the group is a Jordan group. Therefore it seems likely that similar classification theorems can be obtained for slightly weaker definitions of Jordan groups. A start has been made by Adeleke (1994) where he has defined a weaker notion of a *c-Jordan set* and obtained the following classification theorem for them.

Theorem 16.9 (Adeleke 1994) *Let (G, Ω) be an infinite simply primitive permutation group which is c-Jordan. Then G preserves on Ω a dense linear order, a dense upper semilinear order, or a C-relation.*

One can ask the same question for other classes of primitive permutation groups. To cite a specific example, let us make the following definition.

Definition 16.10 Let G be a group acting on an infinite set Ω. A subset Γ of Ω is said to be a *weak Jordan set* if $|\Gamma| > 1$ and for all finite $\Delta \subseteq \Omega \setminus \Gamma$ and for all $\alpha, \beta \in \Gamma$, there exists $g \in G_{(\Delta)}$ with $\alpha^g = \beta$.

PROBLEM 15. Classify primitive closed weak Jordan groups, that is, classify all groups with weak cofinite Jordan sets.

Bibliography

Adeleke, S. A. (1994) A generalisation of Jordan groups, in Richard Kaye & Dugald Macpherson (eds.), *Automorphisms of First-Order Structures,* O. U. P. Oxford (1994), 233–239.

Adeleke, S. A. (1995) Semilinear tower of Steiner systems, *Journal of Combinatorial Theory, Series A* **72** (1995), 243–255.

Adeleke, S. A. (preprint) On irregular infinite Jordan groups, preprint.

Adeleke, S. A. & Macpherson, Dugald (1995) Classification of infinite primitive Jordan permutation groups, *Proceedings of the London Mathematical Society, Series 3* **72** (1995), 63–123.

Adeleke, S. A. & Neumann, Peter M. (1996a) Primitive permutation groups with primitive Jordan sets, *Journal of the London Mathematical Society, Series 2* **53** (1996), 209–229.

Adeleke, S. A. & Neumann, Peter M. (1996b) Infinite bounded permutation groups, *Journal of the London Mathematical Society, Series 2* **53** (1996), 230–242.

Adeleke, S. A. & Neumann, Peter M. (1996c) Relations related to betweenness: their structure and automorphisms, *Memoirs of the Americal Mathematical Society,* to appear.

Baer, R. (1934) Die Kompositionsreihe der Gruppe aller eineindeutigen Abbildungen einer unendlichen Menge auf sich, *Studia Mathematica* **5** (1934), 15–17.

Bertram, E. A. (1973) On a theorem of Schreier and Ulam for countable permutations, *Journal of Algebra* **24** (1973), 316–322.

Bhattacharjee, Meenaxi (1994) Constructing finitely presented infinite nearly simple groups, *Communications in Algebra* **22(11)** (1994), 4561–4589.

Bhattacharjee, Meenaxi (1995) The ubiquity of free subgroups in certain inverse limits of groups, *Journal of Algebra* **172** (1995), 134–146.

Brown, M. (1959) Weak n-homogeneity implies weak $(n-1)$-homogeneity, *Proceedings of the American Mathematical Society* **10** (1959), 644–647.

Cameron, Peter J. (1976) Transitivity of permutation groups on unordered sets, *Mathematische Zeitschrift* **148** (1976), 127–139.

Cameron, Peter J. (1987) Some treelike objects, *Quarterly Journal of Mathematics, Series 2* **38** (1987), 155–183.

Cameron, Peter J. (1990) *Oligomorphic Permutation Groups*, London Mathematical Society Lecture Notes, Vol. 152, Cambridge University Press (1990).

Cameron, Peter J. (1996) Cofinitary permutation groups, *Bulletin of the London Mathematical Society* **28** (1996), 113–140.

Cherlin, G. L. (1988) Homogeneous tournaments revisited, *Geometriae Dedicata* **26** (1988), 231–240.

Cherlin, G. L. (preprint) The classification of countable homogeneous directed graphs and countable homogenous n-tournaments, *Memoirs of the Americal Mathematical Society*, to appear.

Cherlin, G. L., Harrington, L. & Lachlan, A. H. (1985) $\aleph_0$-categorical, $\aleph_0$-stable structures, *Annals of Pure and Applied Logic* **28** (1985), 103–135.

Dixon, John D. & Mortimer, Brian (1996) *Permutation Groups*, Graduate Texts in Mathematics Series, Vol. 163, Springer (1996).

Droste, M. (1985) Structure of partially ordered sets with transitive automorphism groups, *Memoirs of the American Mathematical Society,* Vol. 334, Providence, RI (1985).

Droste, M. & Göbel, R. (1979) On a theorem of Baer, Schreier and Ulam, *Journal of Algebra* **58** (1979), 282–290.

Droste, M. & Macpherson, Dugald (1991) On k-homogeneous posets and graphs, *Journal of Combinatorial Theory, Series A* **56** (1991), 1–15.

Engeler, E. (1959) Äquivalenzklassen von n-Tupeln, *Zeitschrift für Mathematische Logik und Grundlagen der Mathematik* **5** (1959), 340–345.

Erdös, P. & Rényi, A. (1963) 'Asymmetric graphs', *Acta Mathematica Academiae Scientarium Hungaricae* **14** (1963), 295–315.

Erdös, P. & Spencer, J. (1974) *Probabilisitic methods in combinatorics,* Academic Press, New York, (1974).

Evans, D. M. (1986) Homogeneous geometries, *Proceedings of the London Mathematical Society, Series 3* **52** (1986), 305–327.

Evans, D. M. (1994) Examples of $\aleph_0$-categorical structures, in Richard Kaye & Dugald Macpherson (eds.), *Automorphisms of First-Order Structures,* O. U. P. Oxford (1994), 33–72.

Fraïssé, R. (1953) Sur certaines relations qui généralisent l'ordre des nombres rationnels, *Comptes Rendus de l'Académie des Sciences de Paris* **237** (1953), 540–542.

Fraïssé, R. (1986) *Theory of relations,* North-Holland, Amsterdam (1986).

Glass, A. (1981) *Ordered Permutation Groups,* London Mathematical Society Lecture Notes, Vol. 55, Cambridge University Press (1981).

Goode, J. (1989) Hrushovski's geometries, in B. Dahn & H. Wolter (eds.), *Proceedings of the 7th Easter Conference on Model Theory,* (1989), 106–118.

Gorenstein, Daniel (1982) *Finite Simple Groups*, Plenum Press, New York (1982).

Hall, Marshall Jr. (1959) *The theory of groups*, Macmillan, New York (1959).

Hall, Philip (1962) Wreath powers and characteristically simple groups, *Proceedings of the Cambridge Philosophical Society* **58** (1962), 170–184. Reprinted in *Collected works of Philip Hall*, Clarendon Press, Oxford (1988), 609–625.

Henson, C. W. (1972) Countable homogeneous relational structures and $\aleph_0$-categorical theories, *Journal of Symbolic Logic* **37** (1972), 494–500.

Higman, D. G. (1967) Intersection matrices for finite permutation groups, *Journal of Algebra* **6** (1967), 22–42.

Hodges, W. A. (1993) *Model Theory*, Cambridge University Press (1993).

Hodges, W. A. (1997) *A shorter model theory*, Cambridge University Press, Cambridge (1997).

Holland, W. C. (1969) The characterisation of generalised wreath products, *Journal of Algebra* **13** (1969), 152–172.

Hrushovski, E. (1992) Unimodular minimal theories, *The Journal of the London Mathematical Society, Series 2* **46** (1992), 385–396.

Hrushovski, E. (1993) A new strongly minimal set, *Annals of Pure and Applied Logic* **62** (1993), 147–166.

Huntington, E. V. (1935) Inter-relations among the four principal types of order, *Transactions of the American Mathematical Society* **38** (1935), 1–9.

Jordan, Camille (1870) *Traité des substitutions et des équations algébriques*, Gauthier-Villars, Paris (1870).

Jordan, Camille (1871) Théorèmes sur les groupes primitifs, *Journal de Mathématiques Pures et Appliquées* **16** (1871), 383–408. Reprinted in *Oeuvres de Camille Jordan* (ed. J. Dieudonné), Gauthier-Villars, Paris, Vol. 1 (1961), 313–338.

Kantor, W. M. (1985) Homogeneous designs and geometric lattices, *Journal of Combinatorial Theory, Series A* **8** (1985), 64–77.

Lachlan, A. H. (1984) Countable homogeneous tournaments, *Transactions of the American Mathematical Society* **284** (1984), 431–461.

Lachlan, A. H. & Woodrow, R. E. (1980) Countable ultrahomogeneous undirected graphs, *Transactions of the American Mathematical Society* **262** (1980), 51–94.

Luthar, I. S. & Passi, I. B. S. (1996) *Algebra, Volume 1: Group Theory*, Narosa (1996).

Macpherson, Dugald (1985) Orbits of infinite permutation groups, *Proceedings of the London Mathematical Society, Series 3* **51** (1985), 246–284.

Macpherson, Dugald (1994) A survey of Jordan groups, in Richard Kaye & Dugald Macpherson (eds.), *Automorphisms of First-Order Structures*, O. U. P. Oxford (1994), 73–110.

Macpherson, Dugald & Praeger, Cheryl E. (1995) Cycle types in infinite permutation groups, *Journal of Algebra* **175** (1995), 212–240.

McLeish, S. (1994) *The sufficiency of going forth in first order homogeneous structures*, Ph.D. Thesis, Queen Mary and Westfield College, (1994).

Möller, Rögnvaldur G. (1991) The automorphism groups of regular trees, *The Journal of the London Mathematical Society, Series 2* **43** (1991), 236–252.

Möller, Rögnvaldur G. (1992) Ends of Graphs, *Mathematical Proceedings of the Cambridge Philosophical Society* **111** (1992), 255–266.

Neumann, B. H. (1963) Twisted wreath products of groups, *Archiv der Mathematik (Basel)* **14** (1963), 1–6.

Neumann, Peter M. (1964) On the structure of standard wreath products of groups, *Mathematische Zeitschrift* **84** (1964), 343–373.

Neumann, Peter M. (1975) The lawlessness of groups of finitary groups, *Archiv der Mathematik (Basel)* **26** (1975), 561–566.

Neumann, Peter M. (1976) The structure of finitary permutation groups, *Archiv der Mathematik (Basel)* **27** (1976), 3–17.

Neumann, Peter M. (1979) A lemma that is not Burnside's, *The Mathematical Scientist* **4** (1979), 133–141.

Neumann, Peter M. (1985) Some primitive permutation groups, *Proceedings of the London Mathematical Society. Series 3* **50** (1985), 265–281.

Neumann, Peter M. (1996) The classification of some infinite Jordan groups, *Quarterly Journal of Mathematics, Oxford* (2) **47** (1996), 107–121.

Neumann, Peter M. & Praeger, C. E. (1995) Problem 11.70 (p. 83) of *the Kourovka Notebook* titled *Unsolved problems in group theory*, (V. D. Mazurov & E. I. Khukhro eds.) Institute of Mathematics, Siberian Division of the Russian Academy of Sciences. Novosibirsk, (1995).

Neumann, Peter M., Stoy, Gabrielle A. & Thompson, Edward C. (1994) *Groups and Geometry*, O. U. P. Oxford (1994).

Samuel, Pierre (1988) *Projective Geometry*, Undergraduate Texts in Mathematics, Springer-Verlag, New York 1988.

Rado, R. (1964) 'Universal graphs and universal functions', *Acta Arithmetica* **9** (1964), 331–340.

Robinson, Derek J. S. (1995) *A Course in the Theory of Groups*. Graduate Texts in Mathematics Series, Vol. 80, Springer (1995).

Rosenberg, A. (1958) The structure of the infinite general linear group, *Annals of Mathematics* **68** (1958), 278–294.

Rotman, J. J. (1995) *An introduction to the theory of groups*, 4th ed. Graduate Texts in Mathematics Series, Vol. 148, Springer (1995).

Ryll-Nardzewski, C. (1959) On categoricity in power $\leq \aleph_0$, *Bulletin de l'Académie Polonaise des Sciences* **7** (1959), 545–548.

Schmerl, J. H. (1979) Countable homogeneous partially ordered sets, *Algebra Universalis* **9** (1979), 317–321.

Schreier, J. & Ulam, S. (1933) Ueber die Permutationsgruppe der natürlichen Zahlenfolge, *Studia Mathematica* **4** (1933), 134–141.

Suzuki, Michio (1982) *Group Theory I*, Vol. 247, Springer-Verlag (1982).

Suzuki, Michio (1986) *Group Theory II*, Vol. 248, Springer-Verlag (1986).

Svenonius, L. (1959) $\aleph_0$-categoricity in first order predicate calculus, *Theoria* **25** (1959), 82–94.

Thomas, S. R. (1986) Groups acting on infinite dimensional projective spaces, *Journal of the London Mathematical Society, Series 2* **34** (1986), 265–273.

Tits, J. (1952) Sur les groupes doublement transitifs continus, *Commentarii Mathematici Helvetici* **26** (1952), 203–224.

Truss, J. K. (1985) The group of the countable universal graph, *Mathematical Proceedings of the Cambridge Philosophical Society* **98** (1985) 213–245.

Truss, J. K. (1989) Infinite permutation groups II: Subgroups of small index, *Journal of Algebra* **120** (1989), 494–515.

Wagner, Frank O. (1994) Relational structures and dimensions, in Richard Kaye & Dugald Macpherson (eds.), *Automorphisms of First-Order Structures*, O. U. P. Oxford (1994), 153–180.

Wells, C. (1976) Some applications of the wreath product construction, *American Mathematical Society Monthly* **83** (1976), 317–338.

Wielandt, H. (1960) Unendliche Permutationsgruppen, Lecture notes, Universität Tübingen (1960). Reprinted by York University, Toronto, (1967).

Wielandt, H. (1964) *Finite Permutation Groups*. Academic Press, New York (1964). Reprinted in *Mathematische Werke* (ed. Bertram Huppert), Walter de Gruyter, Berlin, Vol. 1 (1994), 119–198.

Wielandt, H. (1967) Endliche k-homogene Permutationsgruppen, *Mathematische Zeitschrift* **101** (1967), 142.

Zil'ber, B. I. (1979) Totally categorical theories: Structural properties and non-finite axiomatizability, in *Model Theory of Algebra and Arithmetic* (ed. L. Pacholski), Lecture Notes in Mathematics, Vol. 834, Springer-Verlag (1979), 381–410.

Zil'ber, B. I. (1984) The structure of models of uncountably categorical theories, *Proceedings of the International Congress of Mathematicians, August 16-24, 1983, Warszawa,* Państwowe Wydawnictwo Naukowe, (1984), 359–368.

List of Symbols

The list is in the order in which the symbols first appear.

$H \leq G$	H is a subgroup of G
$A := B$	A is defined to be B
Ha	the (right) coset of H containing a
$|G : H|$	the index of H in G
G/H, $\cos(G : H)$	the set of cosets of H in G
$A \cup B$	the union of A and B
$|S|$	the number of elements in S
$N \trianglelefteq G$	N is a normal subgroup of G
AB, $A \cdot B$	$\{ab \mid a \in A, b \in B\}$
G/N	the factor group of G by N
$G \cong H$	G is isomorphic to H
$\operatorname{Im} \theta$	the image of θ
$\operatorname{Ker} \theta$	the kernel of θ
$A \cap B$	the intersection of A and B
$\operatorname{Aut}(G)$	the group of automorphisms of G
$C_G(g)$	the centraliser of g in G
$Z(G)$	the centre of G
$\langle A \rangle$	the group generated by A
G'	the derived subgroup of G
$A \times B$	the direct product of A and B

189

$A \rtimes B$	the semidirect product of A by B
$\mathrm{Sym}\,(\Omega)$	the symmetric group on Ω
S_n	the finite symmetric group of degree n
$\mathrm{GL}(2, \mathbf{R})$	the group of 2×2 invertible real matrices
$\mathbf{R}$	the set of real numbers
$\mathbf{R}^2$	2-dimensional Euclidean space
$\Omega^{(k)}$	the set of all k-element ordered subsets of Ω
$\Omega^{\{k\}}$	the set of all k-element unordered subsets of Ω
$\wp(\Omega)$	the power set of Ω
G^Ω	the permutation group induced by G on Ω
$(\alpha_1 \, \alpha_2 \, \ldots \, \alpha_n)$	a finite cycle of length n
$(\ldots \alpha_{-1} \, \alpha_0 \, \alpha_1 \, \ldots)$	an infinite cycle
$\mathbf{Z}$	the set of integers
$\aleph_0, \omega$	the first infinite cardinal
$\mathbf{Q}$	the set of rational numbers
$\mathbf{Q}^+$	the set of positive rational numbers
A_n	the finite alternating group of degree n
α^G	the orbit of α under G
G_α	the stabiliser of α in G
$\mathbf{N}$	the set of natural numbers
$G_{\{\Delta\}}$	the setwise stabiliser of Δ in G
$G_{(\Delta)}$	the pointwise stabiliser of Δ in G
G^Δ	$G_{\{\Delta\}}/G_{(\Delta)}$
$\Omega \setminus \Delta$	the set of elements in Ω not in Δ
$\mathrm{Aut}\,(\mathbf{Q}, <)$	the group of order-automorphisms of $\mathbf{Q}$
$\binom{s}{t}$	s choose t
$\mathrm{fix}_\Omega(g)$	the number of points in Ω that are fixed by g
$\approx, \rho$	an equivalence relation, a G-congruence
$A \Leftrightarrow B$	A is equivalent to B, A if and only if B
$A \Rightarrow B$	A implies B, if A then B

$\alpha \equiv \beta \pmod{\rho}$	α and β are equivalent under ρ
$\rho(\theta)$	the ρ-class containing θ
C_p	the cyclic group of order p
D_∞	the infinite dihedral group
$C_G(A)$	the centraliser of A in G
Δ_0	the diagonal orbital of G
Δ	an orbital of G
$\Delta(\alpha)$	the α-suborbit corresponding to Δ
(V, E)	a graph with vertex set V and edge set E
$(\alpha, \beta)^G$	the orbital of G containing (α, β)
Δ^*	the paired orbital of Δ
$\preceq$	a pre-order
$A \dot\cup B$	the disjoint union of A and B
$\mathrm{supp}(g)$	the support of g
$\mathrm{BS}(\Omega, k)$	the bounded symmetric group on Ω
$\mathrm{BS}(\Omega, \aleph_0), \mathrm{FS}(\Omega)$	the finitary symmetric group on Ω
$\mathrm{Alt}(\Omega)$	the (finitary) alternating group on Ω
$\aleph_1$	the second infinite cardinal
$M_n(F)$	the set of square matrices of degree n
$\mathrm{GL}(n, F)$	the n-dimensional general linear group over the field F
I_n	the unit matrix of degree n
F^n	the n-dimensional product space of a field F
$\mathrm{GL}(V)$	the group of invertible linear transformations over V
$\mathrm{SL}(n, F)$	the n-dimensional special linear group over F
$\det(M)$	the determinant of M
F^*	the non-zero elements in F
$\rho|_A$	the restriction of ρ to A

$GL(n, q)$	the n-dimensional general linear group over the finite field of q elements
$SL(n, q)$	the n-dimensional special linear group over the finite field of q elements
$PGL(n, F)$	the n-dimensional projective general linear group over F
$PSL(n, F)$	the n-dimensional projective special linear group over F
Σ_k	the set of $(k+1)$-dimensional subspaces of a vector space
$PG(V)$	the $(n-1)$-dimensional projective space of V
$\mathcal{P}_n(F)$	the $(n-1)$-dimensional projective geometry over F
$AGL(V)$	the affine general linear group of V
$T(V)$	the set of translations of V
$\mathcal{A}_n(F)$	the n-dimensional affine geometry over F
$AG(V)$	the n-dimensional affine space of V
$\Gamma L(V)$	the semilinear group of V
$P\Gamma L(V)$	the projective semilinear group of V
$A\Gamma L(V)$	the affine semilinear group of V
C^Δ	the (cartesian) product of Δ copies of C
$C \operatorname{Wr} D, C \operatorname{Wr}_\Delta D$	the wreath product of C by D
$C \operatorname{wr} D, C \operatorname{wr}_\Delta D$	the restricted wreath product of C by D
Ω/ρ	the set of ρ-classes in Ω
$C^{(\Delta)}$	the restricted direct product of Δ copies of C
$\mathcal{R}$	a set of finitary relations
$\operatorname{Aut}(\Omega, \mathcal{R})$	the group of automorphisms of Ω preserving the relations in $\mathcal{R}$
M_{22}, M_{23}, M_{24}	the finite Mathieu groups

Homeo(Ω)	the group of homeomorphisms of Ω
$(\Omega, \mathcal{B})$	a Steiner system with points Ω and blocks $\mathcal{B}$
$(\Lambda, \leq)$	a partially ordered set
$\mathcal{B}(\alpha; \beta, \gamma)$	a linear betweenness relation
$\mathcal{K}(\alpha, \beta, \gamma)$	a circular order
$\mathcal{S}(\alpha, \beta; \gamma, \delta)$	a separation relation
$(P, \leq)$	a semilinear order
$C(\alpha; \beta, \gamma)$	a C-relation
$B(x; y, z))$	a B-relation, a general betweenness relation
$D(\alpha, \beta; \gamma, \delta)$	a D-relation
$MJ(\alpha_1, \alpha_2, \ldots, \alpha_n/\omega)$	the unique maximal Jordan set containing ω and omitting $\alpha_1, \alpha_2, \ldots, \alpha_n$
$\mathcal{N} \leq \mathcal{M}$	$\mathcal{N}$ is a substructure of $\mathcal{M}$
Aut$(\mathcal{M})$	the group of automorphisms of a model $\mathcal{M}$
K_n	the complete graph of size n
dom(ϕ)	the domain of ϕ
range(ϕ)	the range of ϕ
$A \hookrightarrow B$	an embedding from A into B
$A \leq_e B$	A is $\mathcal{E}$-embedded in B
Cl(A)	the closure of A
$\delta(H)$	$|V| - |E|$, where H is the hypergraph (V, E)
$\mathcal{H}_{k+1}$	the class of all $(k+1)$-hypergraphs
$\mathcal{A}_k$	a particular subclass of $\mathcal{H}_{k+1}$
$A \leq_{we} B$	A is well-embedded in B
Env$_M(A)$	the envelope of A in M
$d_M(A)$	$\delta(\text{Env}_M(A))$
Cl$_M(A)$	the closure of A in M
$B \oplus_A C$	the free amalgam of B and C over A

Index

General Remarks

Lecture Notes are printed by photo-offset from the master-copy delivered in camera-ready form by the authors. For this purpose Springer-Verlag provides technical instructions for the preparation of manuscripts.

Careful preparation of manuscripts will help keep production time short and ensure a satisfactory appearance of the finished book. The actual production of a Lecture Notes volume normally takes approximately 8 weeks.

Authors receive 50 free copies of their book. No royalty is paid on Lecture Notes volumes.

Authors are entitled to purchase further copies of their book and other Springer mathematics books for their personal use, at a discount of 33,3 % directly from Springer-Verlag.

Commitment to publish is made by letter of intent rather than by signing a formal contract. Springer-Verlag secures the copyright for each volume.

Addresses:

Professor A. Dold
Mathematisches Institut
Universität Heidelberg
Im Neuenheimer Feld 288
D-69120 Heidelberg, Germany

Professor F. Takens
Mathematisch Instituut
Rijksuniversiteit Groningen
Postbus 800
NL-9700 AV Groningen
The Netherlands

Professor Bernard Teissier
École Normale Supérieure
45, rue d'Ulm
F-7500 Paris, France

Springer-Verlag, Mathematics Editorial
Tiergartenstr. 17
D-69121 Heidelberg, Germany
Tel.: *49 (6221) 487-410